NumPy
for
Data Analysis

MADALA HIMA JYOTHI

Assistant Professor
Department of Computer Science and Engineering
Dhanekula Institute of Engineering and Technology
Andhra Pradesh

SUNDARADASU SURESH

Professor and HOD
Department of Computer Science and Engineering
Dhanekula Institute of Engineering and Technology
Andhra Pradesh

PHI Learning Private Limited
Delhi-110092
2025

*In fond memory of **Shri Asoke K. Ghosh** (October 1942 – February 2024), Founder Chairman and Managing Director of PHI Learning, whose vision endlessly inspires.*

The Legacy Continues . . . .

Published by Pushpita Ghosh, PHI Learning Private Limited, Rimjhim House, 111, Patparganj Industrial Estate, Delhi-110092 and Printed by Multi Colour Services, I-45, DLF Industrial Area, Sector-32, Faridabad, Haryana-121003.

₹595.00

NumPy for Data Analysis
Madala Hima Jyothi and Sundaradasu Suresh

ISBN-978-93-91818-35-7 (Print Book)
ISBN-978-93-91818-30-2 (e-Book)

The export rights of the book are vested solely with the publisher.

TO

my beloved father

MADALA YOGESWARA RAO

— Ms. Madala Hima Jyothi

CONTENTS

PREFACE

Python, an easy to understand programming language, is surrounded by vast libraries with add-on modules. It is a trending language in software development. A platform-independent and scripted language, it supports the fusion of complex solutions for pre-built components.

Why should one learn Python?

- Python is a freeware and open-source application, so we can customize Python as we like.
- Python is a platform-independent programming language.
- Python is a simple object-oriented programming language. It had simple syntax and code compared to all other programming languages.
- We can develop various desktop, standalone, scripting applications easily using Python.

Python is the programming language used for data science as it contains tools for performing mathematical and statistical operations. This is the reason why data scientists use Python. The major part of data science is involved with data preprocessing and cleaning, which is time-consuming. NumPy is the interface used to store and operate on dense data.

NumPy is a Python library. Travis Oliphant created it in 2005. NumPy aims to provide an array object which is 50 times faster than a traditional Python list. NumPy is a Python library that is a freeware and open-source library. It is very fast as most of NumPy is written in C language. NumPy can handle large datasets effectively and efficiently.

NumPy is a viral Python module. It is heavily used in scientific computing. As compared to Python list, NumPy array is fast, convenient and takes less memory than list. It is the backbone for the remaining Python libraries such as Pandas, Matplotlib, Scilearn, etc. NumPy library is used for complex mathematical functions. It is the core library for scientific calculations, which contains a basic n-dimensional array object.

Organization of the Book

NumPy for Data Analysis is a practical and beginner-friendly introduction to data analysis covering the basics of NumPy (Numerical Python), a data science tool in Python.

Chapter 1 discusses the basics of Python, various Python libraries, and the history of NumPy with installation. Chapter 2 describes the creation of nd arrays and accessing the elements of the NumPy array using indexing and slicing. Chapter 3 deals with changing the data type of an existing array, creating a reference to an existing array, and creating an external copy of data object and string operations for arrays. Chapter 4 covers accessing the elements of nd array using indexing, slicing, accessing multiple elements, which are not in order, using advanced indexing, and selecting elements of an array based on condition. Chapter 5 explains how to perform various arithmetic functions between nd array and ufunc. Chapter 6 reviews that the broadcasting will be performed internally while performing arithmetic operations. Chapter 7 includes the array manipulation functions like reshape(), resize(), ravel(), flat variable, transpose() and swapaxes(). Chapter 8 tells about how to concatenate multiple nd arrays into single array using stack(), vstack(), hstack(), dstack() and to perform split operation on nd array using split(), vsplit(), hsplit(), dsplit(). Chapter 9 deals with the sorting elements of nd array using sort() function, search elements in nd array using where(), and order keywords to specify the field in sorting a structured array. Chapter 10 describes inserting or deleting elements in an array at the required position using functions. Chapter 11 explains how to create nd array using class matrix in NumPy library and perform matrix level multiplication using dot(). Chapter 12 discusses how to perform linear algebra operations using the linalg function and find the determinant of a matrix using det(). How to find the unique items and their count in nd array using unique() function is covered in Chapter 13. Chapter 14 is devoted to collecting, storing, and analyzing huge amount of data using basic statistic functions. Chapter 15 discusses the mathematical and financial operations using NumPy libraries. Finally, Chapter 16 describes the basic concepts of functional programming.

We would like to thank our Chairman Shri Dhanekula Ravindranath Tagore, Secretary Shri Dhanekulu Bhavani Prasad, Director Shri D K R K Ravi Prasad, and Principal Dr. Ravi Kadiyala for their support in completing this book successfully. We express our sincere thanks to our family members who inspire us to do our best.

The first author is grateful to her husband, Rakesh Chowdary Gutta, children, Akshaya and Abhiram, mother and brother, for their love, support, and encouragement throughout this book writing project.

We are thankful to our friends and colleagues, especially Ms. L.N.B. Jyotsna and Ms. N. Madhuri, for reading the script at various stages and providing valuable feedback to improve the book's coverage.

Though we have put in all our efforts to make this book student friendly, we welcome suggestions/feedback from the students and faculty members to improve the content and coverage of the book.

Madala Hima Jyothi
Sundaradasu Suresh

1 INTRODUCTION

1.1 WHAT IS DATA SCIENCE?

Data science is a field of study that deals with vast volume of data using modern tools to derive meaningful information. In simple words, data science is about finding data within data in the real world and solving real-world problems by using this data. There are various factors to be considered to decide which language should be preferably used to analyze the data:

1. Speed
2. Packages
3. Design goals

After considering these factors, one can infer that Python is a perfect fit for data science roles.

1.2 WHY PYTHON?

Python was initially designed by Guido Van Rossum in the late 1980s. After some constant improvements and a lot of bug fixes, it was officially launched in 1989. Python is a widely used general-purpose, high-level programming language. Even though Python is developed under an open source license, it can be used for commercial purposes as well.

The Python language is one of the widely used programming languages because of its simplified syntax. Due to its ease of learning, Python codes can be easily written and executed faster than any other programming languages.

Python has plenty of data processing tools which help in better handling of data. Python is important for data scientists, as it provides a vast variety of applications used in data science. It has a lot of packages like TensorFlow, Kara's Tiano which help data scientists to develop trending Deep Learning algorithms quickly. In Python, approximately 72000 packages are available. Python has a package index, sometimes called PyPi. It is in open source so anyone can create a library.

1.3   WHAT IS IDE?

Many of us prefer a text editor for writing the code and then open a terminal window for executing the code. When an error has been detected, again we have to switch back to the text editor and need to make the necessary changes. After that, we need to run the code again. Typically, for large-scale problems, we need an efficient testing module where we can write, run and play with the code, that too in one place. Integrated Development Environment (IDE) provides not only the capability of coding, but also code testing, running code, highlighting syntax, bracket matching, auto-completion, debugging code, code suggestion, and various other features. There are several IDEs for Python:

1. Jupyter Notebook
2. PyCharm
3. Spyder
4. Canopy
5. VIM
6. ATOM

Jupyter Notebook

Jupyter was introduced in the year 2014 after its predecessor iPython. Python has redefined the concepts of an IDE and is adding more features to it. It is a web application-based server-client structure, which is easy to use and to create, analyze and manipulate documents. Since it is a web interface, it may integrate with many of the existing web libraries. It is not just an IDE but it is also widely used as an educational tool for presentation and even for writing blogs.

PyCharm

It was invented by the company JetBrains and other IDs like Java ID. It is much better for working with multiple scripts, handling multiple files, and linking them together. It supports Anaconda and includes packages like NumPy, Matplotlib and so on. Just like other IDE, it has a powerful debugger with a graphical user interface. Python allows you to add plugins for nonpythonic files.

These plugins handle indentation and highlight the errors and keywords. It is both professional and community edition and it is not open source completely. The main drawback is that it is memory intensive.

Spyder

Spyder (Scientific Python Development Environment) is a free and open source scientific environment written in Python that is included with Anaconda. It was built for the data science community which is integrated with essential data-centric libraries like NumPy, Matplotlib, Pandas, iPython, and so on. The built-in capabilities can be extended further by plugins and API's. Spyder contains features like a text editor with syntax highlighting, code completion, static code analysis, debugging and variable exploring.

1.4   PYTHON LIBRARIES FOR DATA ANALYSIS

Python has extraordinary Python libraries for data science that made Python popular among data scientists in solving problems. Some of them are as follows:

1. NumPy
2. Pandas
3. SciPy
4. Matplotlib
5. Scikit Learn

NumPy

NumPy is the fundamental package for scientific computing with Python.

Features

- creates powerful n-dimensional array
- contains tools for integrating C/C++ and Fortran code
- has useful linear algebra, Fourier Transform, and random number capabilities

FIGURE 1.1   NumPy

Pandas

Pandas is used for structured data operations and manipulations.

Features

- most useful Data Analysis library in Python
- instrumental in increasing the use of Python in the Data Science community
- extensively used for data munging and preparation

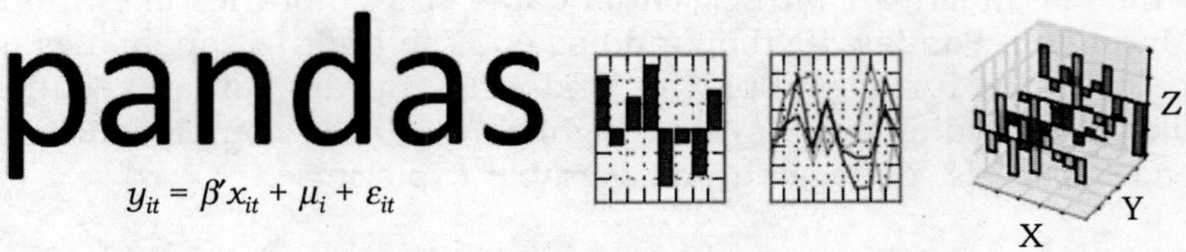

FIGURE 1.2 Pandas

SciPy

SciPy (Scientific Python) is used for high-level computations.

Features

- has a set of scientific and numerical tools for Python
- currently supports special functions, integration, Ordinary Differential Equation (ODE) solvers, gradient optimization, and others
- has fully featured versions of the linear algebra modules.

FIGURE 1.3 SciPy

MatPlotlib

It is a primary leaf for visualization purposes. It has a very powerful feature of visualizing data, i.e., exploratory data analysis for doing the univariate and bivariate analysis.

FIGURE 1.4 MatPlotlib

Scikit Learn

It is useful for performing Machine Learning activities like linear regression, classification, and so on.

FIGURE 1.5   Scikit Learn

Besides this, there are some of the following additional libraries too:

1. Network and I graph
2. TensorFlow
3. BeautifulSoup
4. OS
5. Scrappy
6. Plotly
7. Seaborn
8. PyCaret
9. Keras
10. PyTorch

1.5   HISTORY OF NUMPY

NumPy which stands for Numerical Python, is the fundamental Python library to perform complex numerical operations. Jim Hugunin created a package called Numeric in 1995 which is considered as the parent package of NumPy. Later, Travis Oliphant developed NumPy in the year 2005.

FIGURE 1.6   Jim Hugunin

NumPy is a freeware and open-source library which is written in C and Python. Most of NumPy is written in C, so performance-wise NumPy is the best like C. Because of high speed, NumPy is the best choice for Machine Learning algorithms than traditional Python's in-built data structures like List. But in Data Science, Machine Learning, Deep Learning and Artificial Intelligence require complex mathematical operations such as:

1. Creation of arrays
2. Perform several operations on those arrays
3. Differential equations
4. Statistics related operations
5. Integral calculus operations

The above type of operations are not supported by Python but Numpy can perform all these operations.

FIGURE 1.7   Travis Oliphant

1.5.1   Importance of NumPy

1. NumPy has Vectorization features.
2. nd array (n-dimensional array) is the basic data structure in NumPy.
3. Numpy is the backbone of remaining libraries like Pandas, Matplotlib, Sklearn(sci-kit learn), etc.

1.5.2   Features of NumPy

1. It is fast because most of NumPy is written in C language.
2. Numpy Array (nd array) is treated as a basic data structure in Numpy.

3. Numpy has a vectorization feature which improves performance while iterating elements.

1.5.3 Applications of NumPy

NumPy is used to:

1. perform linear algebra functions
2. perform linear regression
3. do K-means clustering
4. make control systems
5. do operational research
6. perform logistic regression

1.6 ARRAY

A row of homogenous data elements is nothing but an array. It is the most commonly used concept in programming languages like C/C++, Java, etc. By default, array's concept is not available in Python, instead, we can use List.

Note: In Python, we can create arrays in the following two ways:

- By using array module
- By using NumPy module

1.6.1 nd Array in NumPy

1. In NumPy, data is stored in the form of an array using Array Data Structure.
2. nd array ==> N-dimensional array or NumPy array
3. This nd array is most commonly used in Data Science libraries like Pandas, Scikit learn, etc.

1.7 INSTALLING NUMPY MODULE

To install the latest version,

```
pip install numpy
```

To install a particular version of NumPy,

```
pip install numpy==1.20.0
```

Note: If you install Anaconda Python, then there is no need to install NumPy separately. NumPy module is implicitly installed along with Anaconda Python.

To check the version of Numpy installed,

```
import numpy as np9
print(np9.__version__)
```

1.8   PYTHON LIST VS NUMPY ARRAY

Similarities

1. Both can be used to store data.
2. The order will be preserved in both. Hence indexing and slicing concepts are applicable.
3. Both are mutable, i.e., we can change the content.

Differences

1. The Python list is inbuilt. However, we have to install and import NumPy explicitly.
2. The list can contain heterogeneous elements. But array contains only homogeneous elements.
3. On the list, we cannot perform vector operations. But on nd array, we can perform vector operations.
4. Arrays consume less memory than a list.
5. Arrays are super fast as compared to the list.
6. NumPy arrays are more convenient to use while performing complex mathematical operations.

PRACTICE PROGRAMS

Program 1: A program that is on the list where we cannot perform vector operations. But on nd array we can perform vector operations.

```
import numpy as np9
h = [10,20,30,40]
h+2
```

OUTPUT

```
C:\book>py h.py
Traceback (most recent call last):
    File "C:\book\h.py", line 4, in <module>
    1+9
TypeError: can only concatenate list (not "int") to list
```

On ndarrays, we can perform vector operations

```
import numpy as np9
h = np9.array([10,20,30,40])
print(h+9)
```

OUTPUT

```
C:\book>py h.py
[19      29      39      49]
```

Program 2: A program to check that arrays consume less memory than a list.

```
import numpy as np9
import sys
l=[10,20,30,40,50,60,70,80,90,100,10,20,30,40,50,60,70,80,90,100]
a=np9.array([10,20,30,40,50,60,70,80,90,100,10,20,30,40,
50,60,70,80,90,100])
print('The Size of list l : ",sys.getsizeof(l))
print('The Size of ndarray a : ',sys.getsizeof(a))
```

OUTPUT

```
C:\book>py h.py
The Size of list l :   216
The Size of ndarray a :   176
```

Program 3: A program to check that arrays are super fast as compared to a list.

```
import numpy as np9
from datetime import datetime
a = np9.array([10,20,30])
b = np9.array([1,2,3])
#dot product: A.B=10X1 + 20X2 +30X3=140
#traditional python code
def dot_product(a,b):
    result = 0
    for i,j in zip(a,b):
        result = result + i*j
    return result
before = datetime.now()
for i in range(1000000):
    dot_product(a,b)
after = datetime.now()
print('The Time taken by traditonal python:',after-before)
#numpy library code
before = datetime.now()
for i in range(1000000):
    np9.dot(a,b) # this is from numpy
after = datetime.now()
print('The Time taken by Numpy Library:',after-before)
```

OUTPUT

```
C:\book>py h.py
The Time taken by traditonal python: 0:00:04.591271
The Time taken by Numpy Library: 0:00:02.532104
```

EXERCISES

1. Write a program to check the time taken to add elements in Python list and NumPy array.

2 CREATION OF NUMPY ARRAYS

An array is a data structure that contains a collection of elements in order to create an array in Python from a list or tuples. A value stored in array holds an address which is known as index. It contains an ordered series of elements which means that every value is present at its particular address and this address is specified by its index value.

Note:

- Array module is not recommended to use because the library support is not available.

2.1 CREATION OF ARRAY (nd ARRAY)

NumPy library contains several functions to create nd array based on the requirement. The following are few of such functions:

1. array()
2. arange()
3. linspace()
4. zeros()
5. ones()
6. full()
7. eye()
8. identity()
9. diag()
10. empty()
11. np9.random module
    (i) randint()
    (ii) rand()

    (iii) uniform()
    (iv) randn()
    (v) normal()
    (vi) shuffle()

Note:

- np9.random module is developed based on the array.
- The performance of this module is more when compared to the normal python random module.

1. array()

To create a 1-D array(Vector) creation using list or tuple.

Syntax

```
array(object,dtype=None,copy=True,order='k',subok=False,ndim=0)
```

Array Attributes

The following are various array attributes:

1. ndim: To get the dimension of the array
2. shape: To get a shape of the array
3. size: To get the size of the array which is nothing but a number of elements.
4. dtype: To get the data type of array elements
5. item size: The size of each array element in bytes

Program 1: A program for the creation of 1-D array using a list or tuple.

```
import numpy as np9
l = [60,20,30]
print(f'Type of l --> {type(l)}')
a = np9.array(l)
print(f'Type of a --> {type(a)}')
print(f"a --> {a}")
```

OUTPUT

```
C:\book>py h.py
21Type of l --> <class 'list'>
Type of a --> <class 'numpy.ndarray'>
a --> [60 20 30]
```

2. arange()

We can create only 1-D arrays with arange() function. In Python, creation of range() function is as follows:

1. range(n) → n values from 0 to n-1
   >>>range(4) → 0,1,2,3
2. range(m,n) → from m to n-1
   >>>range(2,7) → 2,3,4,5,6

3. range(begin,end,step)
   >>>range(1,11,1) → 1,2,3,4,5,6,7,8,9,10
   >>>range(1,11,2) → 1,3,5,7,9
   >>>range(1,11,3) → 1,4,7,10

Program 2: A program for printing 1-D array using NumPy (stop value ==> 0 to end-1).

```
import numpy as np9
a = np9.arange(10)
print('a ',a)
print('The dimensions of the array a ',a.ndim)
print('The data type of elements of array ',a.dtype)
print('The size of the array a ', a.size)
print('The shape of the array a', a.shape)
```

OUTPUT

```
C:\book>py h.py
a [0 1 2 3 4 5 6 7 8 9]
The dimensions of the array a 1
The data type of elements of array int32
The size of the array a 10
The shape of the array a (10,)
```

Program 3: A program for printing 1-D array using NumPy (both start and end values: start to end-1).

```
import numpy as np9
a = np9.arange(1,11)
print('a ',a)
print('The dimensions of the array a ',a.ndim)
print('The data type of elements of array a ',a.dtype)
print('The size of the array a ',a.size)
print('The shape of the array a ',a.shape)
```

OUTPUT

```
C:\book>py h.py
a [ 1  2  3  4  5  6  7  8  9 10]
The dimensions of the array a 1
The data type of elements of array a int32
The size of the array a 10
The shape of the array a (10,)
```

Program 4: A program for printing 1-D array with a particular data type of the elements using NumPy.

```
import numpy as np9
a = np9.arange(1,11,3,dtype=float)
print('a ',a)
print('The dimensions of the array a ',a.ndim)
print('The data type of elements of array a ',a.dtype)
print('The size of the array a ',a.size)
print('The shape of the array a ',a.shape)
```

OUTPUT

```
C:\book>py h.py
a [ 1.  4.  7.  10.]
```

```
The dimensions of the array a 1
The data type of elements of array a float64
The size of the array a 4
The shape of the array a (4,)
```

3. linspace()

This function is same as arange() but returns linearly spaced values in the specified interval.

Syntax

```
linspace(start, stop, num=50, endpoint=True, retstep=False,
dtype=None, axis=0)
```

- Both start and stop are included because endpoint=True.
- If endpoint=False then stop is excluded.
- retstep denotes the spacing between the points. If True then the value is returned.
- Calculation of spacing (stop-start)/(num-1) if endpoint=True.
- Calculation of spacing (stop-start)/(num) if endpoint=False.

Program 5: A program that 50 evenly spaced values between 0 and 1 are returned including both 0 and 1.

```
import numpy as np9
print(np9.linspace(0,1))
```

OUTPUT

```
C:\book>py h.py
[0.         0.02040816 0.04081633 0.06122449 0.08163265 0.10204082
 0.12244898 0.14285714 0.16326531 0.18367347 0.20408163 0.2244898
 0.24489796 0.26530612 0.28571429 0.30612245 0.32653061 0.34693878
 0.36734694 0.3877551  0.40816327 0.42857143 0.44897959 0.46938776
 0.48979592 0.51020408 0.53061224 0.55102041 0.57142857 0.59183673
 0.6122449  0.63265306 0.65306122 0.67346939 0.69387755 0.71428571
 0.73469388 0.75510204 0.7755102  0.79591837 0.81632653 0.83673469
 0.85714286 0.87755102 0.89795918 0.91836735 0.93877551 0.95918367
 0.97959184 1.        ]
```

Program 6: A program to print 4 evenly spaced values between 0 and 1 including 0 and 1.

```
import numpy as np9
print(np9.linspace(0,1,4))
```

OUTPUT

```
C:\book>py h.py
[0.         0.33333333 0.66666667 1.        ]
```

Program 7: A program to print 4 evenly spaced values between 0 and 1 including 0 and excluding 1.

```
import numpy as np9
print(np9.linspace(0,1,4,endpoint=False))
```

OUTPUT

```
C:\book>py h.py
[0. 0.25 0.5 0.75]
```

arange() vs linspace()

Table 2.1 arange() vs linspace()

arange()	linspace()
Elements will be considered in the given range based on step value.	The specified number of values will be considered in the given range.

4. zeros()

The array is filled with 0s which returns a new array of given shape and type.

Syntax

```
zeros(...)
zeros(shape, dtype=float, order='C', *, like=None)
    Return a new array of given shape and type, filled with zeros.
```

Note:

- In above syntax, shape will always be created in tuple form where
  - 0-D array ==> Scalar :: single value
  - 1-D array ==> Vector :: collection of 0-D arrays
  - 2-D array ==> Matrix :: collection of 1-D arrays
  - 3-D array ==> Collection of 2-D arrays

Example

1. (10,) ==>1-D array contains 10 elements # Single element in tuple
2. (5,2) ==>2-D array contains 5 rows and 2 columns
3. (2,3,4) ==>3-D array

   where

         2 ==> Number of 2-D arrays
         3 ==> The number of rows in every 2-D array
         4 ==> The number of columns in every 2-D array
   size: 2 * 3 * 4 = 24

Note: In the above example, it creates two 2-D arrays with size 3×4.

Program 8: A program to print 1-D array with zeros.

```
import numpy as np9
print(np9.zeros(3))
```

OUTPUT

```
C:\book>py  h.py
[0. 0. 0.]
```

Program 9: A program to print 2-D arrays with zeros.

```
import numpy as np9
print(np9.zeros((4,3)))
```

OUTPUT

```
C:\book>py  h.py
[[0.  0.  0.]
 [0.  0.  0.]
 [0.  0.  0.]
 [0.  0.  0.]]
```

Program 10: A program to print 3-D arrays with zeros.

```
import numpy as np9
print(np9.zeros((2,3,4)))
```

OUTPUT

```
C:\book>py  h.py
[[[0.  0.  0.  0.]
  [0.  0.  0.  0.]
  [0.  0.  0.  0.]]

 [[0.  0.  0.  0.]
  [0.  0.  0.  0.]
  [0.  0.  0.  0.]]]
```

Note:

- From the above output, we observe that:
    1. In the outermost, there are 3 square brackets that show it is 3-D array.
    2. If one outermost bracket is excluded then it is a 2-D array ==> total two 2-D arrays present.
    3. Each 2-D array contains 1-D arrays. Each 2-D contains 3-rows and 4-columns.

Program 11: A program to print 4-D array with zeros. (collection of 3-D arrays)

```
#  (2,2,3,4)
#  2  ==>  no. of 3-D arrays
#  2  ==>  every 3-D 2 2-D arrays
#  3  ==>  Every 2-D array contains 3 rows
#  4  ==>  every 2-D array contains 4 columns
import numpy as np9
print(np9.zeros((2,2,3,4)))
```

OUTPUT

```
C:\book>py  h.py
[[[[0.  0.  0.  0.]
   [0.  0.  0.  0.]
   [0.  0.  0.  0.]]

  [[0.  0.  0.  0.]
   [0.  0.  0.  0.]
   [0.  0.  0.  0.]]]

 [[[0.  0.  0.  0.]
   [0.  0.  0.  0.]
   [0.  0.  0.  0.]]
```

```
[[0. 0. 0. 0.]
 [0. 0. 0. 0.]
 [0. 0. 0. 0.]]]]
```

5. ones()

It is exactly same as zeros except array is filled with 1 which returns a new array of given shape and type.

Syntax

```
ones(shape, dtype=None, order='C', *, like=None)
    Return a new array of given shape and type, filled with ones.
```

Program 12: A program to print 1-D array using ones().

```
import numpy as np9
print(np9.ones(10))
```

OUTPUT

```
C:\book>py h.py
[1. 1. 1. 1. 1. 1. 1. 1. 1. 1.]
```

Program 13: A program to print 2-D array using ones().

```
import numpy as np9
print(np9.ones((5,2),dtype=int))
```

OUTPUT

```
C:\book>py h.py
[[1 1]
 [1 1]
 [1 1]
 [1 1]
 [1 1]]
```

Program 14: A program to print 3-D array using ones().

```
import numpy as np9
print(np9.ones((2,3,4),dtype=int))
```

OUTPUT

```
C:\book>py h.py
[[[1 1 1 1]
  [1 1 1 1]
  [1 1 1 1]]

 [[1 1 1 1]
  [1 1 1 1]
  [1 1 1 1]]]
```

6. full()

It returns a new array of given shape and type, filled with fill_value.

Syntax

```
full(shape, fill_value, dtype=None, order='C', *, like=None)
    Return a new array of given shape and type, filled with
    'fill_value'.
```

Program 15: A program to print 1-D array using full() function.

```
import numpy as np9
print(np9.full(10,fill_value=2))
```

OUTPUT

```
C:\book>py  h.py
[2 2 2 2 2 2 2 2 2 2]
```

Program 16: A program to print 2-D array using full() function.

```
import numpy as np9
print(np9.full((2,3),fill_value=3))
```

OUTPUT

```
C:\book>py  h.py
[[3 3 3]
 [3 3 3]]
```

Program 17: A program to print 3-D array using full() function.

```
import numpy as np9
print(np9.full((2,3,4),fill_value=8))
```

OUTPUT

```
C:\book>py  h.py
[[[8 8 8 8]
  [8 8 8 8]
  [8 8 8 8]]

 [[8 8 8 8]
  [8 8 8 8]
  [8 8 8 8]]]
```

Note:

- All of the following will give the same results.

```
np9.full(shape=(2,3,4),fill_value=8)
np9.full((2,3,4),fill_value=8)
np9.full((2,3,4),8)
```

7. eye()

It is used to generate identity matrix and returns 2-D array with ones on the diagonal and zeros elsewhere.

Syntax

```
eye(N, M=None, k=0, dtype=<class 'float'>, order='C', *, like=None)
    Return a 2-D array with ones on the diagonal and zeros
    elsewhere.
```

Positional arguments only and keyword arguments only

1. f(a,b) ==> we can pass positional as well as keyword arguments
2. f(a,/,b) ==> before '/', we should pass positional arguments only for variables, i.e., here for a
3. f(a,*,b) ==> after '*', we should pass keyword arguments only for the variables, i.e., here for b

Program 18: A program to print both positional and keyword arguments.

```
def f(x,y):
    print('The value of x',x)
    print('The value of y',y)
f(10,20)
f(x=10,y=20)
```

OUTPUT

```
C:\book>py h.py
The value of x 10
The value of y 20
The value of x 10
The value of y 20
```

Program 19: A program to print the position_only

```
def f1(a, /,b):
    print(f'The value of a :{a}')
    print(f'The value of b :{b}')
    print("Calling f1(10,20)==> Valid")
f1(10,20) # valid
print("Calling f1(a=10,b=20) ==> Invalid")
f1(a=10,b=20)
```

OUTPUT

```
C:\book>py h.py
The value of a :10
The value of b :20
Calling f1(10,20)==> Valid
Calling f1(a=10,b=20) ==> Invalid
Traceback (most recent call last):
    File "C:\book\h.py", line 8, in <module>
        f1(a=10,b=20)
TypeError: f1() got some positional-only arguments passed as
keyword arguments: 'a'
```

Note:

- It is invalid because before '/' the value for variable should be passed as positional only.

Program 20: A program to print the positional arguments.

```
def f1(a,*,b):
    print(f'The value of a :: {a}')
    print(f'The value of b :: {b}')
f1(a=10,b=50) #valid
f1(10,20)
```

OUTPUT

```
C:\book>py h.py
The value of a :: 10
The value of b :: 50
Traceback (most recent call last):
    File "C:\book\h.py", line 5, in <module>
        f1(10,20)
TypeError: f1() takes 1 positional argument but 2 were given
```

Note:

- It is invalid because after "*" we have to provide value for variable(b) through keyword only.

Example: Return 2-D array with ones on the diagonal and zeros elsewhere.

eye() properties:

1. It will always return 2-D arrays.
2. The number of rows and number of columns need not be the same.
3. If we omit the 'M' value then the value will be same as 'N'.
4. By default, main diagonal contains 1s. But we can customize the diagonal which should contain 1s.

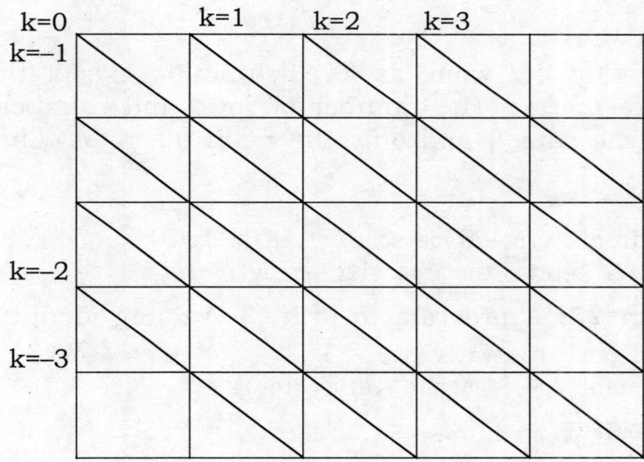

FIGURE 2.1   2-D eye() properties

- N ==> No of rows
- M ==> No. of columns
- K ==> Decides in which diagonal the value should be filled with 1s

        0 ==> Main diagonal
        1 ==> Above the main diagonal
        -1 ==> Below the main diagonal

Program 21: A program to print 2-D array using eye() function with 0 diagonal.

```
import numpy as np9
print(np9.eye(2,3))
```

OUTPUT

```
C:\book>py h.py
[[1. 0. 0.]
 [0. 1. 0.]]
```

Note:

- K → decides in which diagonal the value should be filled with 1s.

Program 22: A program to print 5-D array using eye() function with -1 diagonal.

```
import numpy as np9
print(np9.eye(5,k=-1))
```

OUTPUT
```
C:\book>py h.py
[[0. 1. 0. 0. 0.]
 [0. 0. 1. 0. 0.]
 [0. 0. 0. 1. 0.]
 [0. 0. 0. 0. 1.]
 [0. 0. 0. 0. 0.]]
```

8. identity()

It is exactly the same as 'eye()' function except that it is always a square matrix (The number of rows and number of columns is always the same), and only the main diagonal contains 1s.

Syntax

```
identity(n, dtype=None, *, like=None)
    Return the identity array.
```

Program 23: A program to print 3-D using identity() function.

```
import numpy as np9
print(np9.identity(3,dtype=int))
```

OUTPUT
```
C:\book>py h.py
[[1 0 0]
 [0 1 0]
 [0 0 1]]
```

9. diag()

- diag(v, k=0) ==> Extract a diagonal or construct a diagonal array.
- If we provide 2-D array then it will extract the elements at k-th diagonal.
- If we provide 1-D array, it will construct a diagonal array with the provided elements and the remaining elements are filled with zeros.

Syntax

```
diag(v, k=0)
```

Extract a diagonal or construct a diagonal array.

Program 24: A program to extract 2-D diagonal elements.

```
import numpy as np9
a = np9.arange(1,10).reshape(3,3)
print("Original 2-D array ",a)
print("Elements prsent at 0-diagonal ",np9.diag(a,k=0))
print("Elements prsent at 1-diagonal ",np9.diag(a,k=1))
print("Elements prsent at 2-diagonal ",np9.diag(a,k=2))
print("Elements prsent at -1-diagonal ",np9.diag(a,k=-1))
print("Elements prsent at -2-diagonal ",np9.diag(a,k=-2))
print("Elements prsent at 3-diagonal ", np9.diag(a,k=3))
```

OUTPUT

```
C:\book>py h.py
Original 2-D array [[1 2 3]
                   [4 5 6]
                   [7 8 9]]
Elements prsent at 0-diagonal   [1 5 9]
Elements prsent at 1-diagonal   [2 6]
Elements prsent at 2-diagonal   [3]
Elements prsent at -1-diagonal  [4 8]
Elements prsent at -2-diagonal  [7]
Elements prsent at 3-diagonal   []
```

Program 25: A program to construct 2-D diagonal array with the provided elements and remaining elements are filled with zeros.

```
import numpy as np9
a = np9.array([10,20,30,40])
print(np9.diag(a,k=0))
```

OUTPUT

```
C:\book>py h.py
[[10  0  0  0]
 [ 0 20  0  0]
 [ 0  0 30  0]
 [ 0  0  0 40]]
```

Program 26: A program to construct 11 diagonal for a given 2-D array by providing two arguments (the first argument represents no. of elements and second element represents the diagonal) and the remaining elements are filled with zeros.

```
import numpy as np9
a = np9.array([10,20,30,40])
print(np9.diag(a,k=-1))
```

OUTPUT

```
C:\book>py h.py
[[ 0  0  0  0 0]
 [10  0  0  0 0]
 [ 0 20  0  0 0]
 [ 0  0 30  0 0]
 [ 0  0  0 40 0]]
```

10. empty()

It returns a new array of given shapes and types, without initializing entries.

Syntax

```
empty(shape, dtype=float, order='C', *, like=None)
```

Program 27: A program to print 2-D array of 3×3 matrices using empty() function.

```
import numpy as np9
print(np9.empty((3,3)))
```

OUTPUT

```
array([[0.00000000e+000,  0.00000000e+000,  0.00000000e+000],
 [0.00000000e+000,   0.00000000e+000,   6.42285340e-321],
 [4.47593775e-091,  4.47593775e-091,  1.06644825e+092]])
```

Note:
- There is no default value in the C language.
- If a value is initialized then it will be assigned with some garbage values.
- Generally 'empty()' function is used to create a dummy array.
- np9.empty(10) returns uninitialized data ==> we can use this for future purpose.

zeros() vs empty()

Table 2.2   zeros() vs empty()

zeros()	empty()
1. If we require an array only with zeros then we should go for zeros().	1. If we require an empty array for future purpose, then we should go for empty().
2. The time required to create zero array is more as compared to empty array.	2. It is marginally faster.

Program 28: A program to find the performance comparison of zeros() and empty() functions.

```
import numpy as np9
from datetime import datetime
import sys
begin = datetime.now()
a = np9.zeros((10000,300,400))
after = datetime.now()
print('Time taken by zeros:',after-begin)
a= None
begin = datetime.now()
```

```
a = np9.empty((10000,300,400))
after = datetime.now()
print('Time taken by empty:',after-begin)
OUTPUT
C:\book>py h.py
Time taken by zeros: 0:00:00.199965
Time taken by empty: 0:00:00.157669
```

Note:

- NumPy: basic data type ==> ndarray
- Scipy: advance NumPy
- Pandas: basic data types => Series(1-D) and DataFrame (multi dimensional) based on NumPy
- Matplotlib: data visualization module
- Seaborn, plotly: based on Matplotlib

2.2 np9.random MODULE

This library contains several functions to create nd arrays with random data.

   (i)   randint() ==> to generate random int values in the given range
  (ii) rand() ==> uniform distribution in the range [0,1)
 (iii) uniform() ==> uniform distribution in the provided range
 (iv) randn() ==> normal distribution values with mean 0 and variance 1
  (v) normal() ==> normal distribution values with mean and variance.

(i) randint: To generate random integer values in the given range.

Synatax

```
randint(low, high=None, size=None, dtype=int)
```

- Return random integers from low (inclusive) to high (exclusive).
- It is represented as [low, high) ==> '[' means inclusive and ')' means exclusive.
- If high is None (the default), then results are from [0, low).

Program 29: A program to generate a single random integer value in the range 10 to 19.

```
import numpy as np9
print(np9.random.randint(10,20))
OUTPUT
C:\book>py h.py
15
```

Program 30: A program to create 1-D nd array of size 10 with random values from 1 to 8.

```
import numpy as np9
print(np9.random.randint(1,9,size=10))
```

OUTPUT

```
C:\book>py h.py
[3 8 5 4 6 5 2 2 8 3]
```

Program 31: A program to create 2-D array with shape (3,5).

```
import numpy as np9
print(np9.random.randint(100,size=(3,5)))
```

OUTPUT

```
C:\book>py h.py
[[15 36 9 46 69]
 [27 95 83 38 15]
 [53 26 43 81 85]]
```

Program 32: A program to print random values from 0 to 99 using 2-D array.

```
import numpy as np9
print(np9.random.randint(100,size=(3,5)))
```

OUTPUT

```
C:\book>py h.py
[[23 26 87 14 42]
 [33 76 9 51 45]
 [52 91 43 7 1]]
```

Program 33: A program to create 3-D array with shape (2,3,4). 3-D array contains two 2-D arrays, each 2-D array contains 3 rows and 4 columns

```
import numpy as np9
print(np9.random.randint(100,size=(2,3,4)))
```

OUTPUT

```
C:\book>py h.py
[[[68 23 33 99]
  [ 3  1  4 29]
  [35 12 12 33]]

 [[33 77  5 44]
  [36 71 29 80]
  [62 62 27 41]]]
```

Uniform Distribution vs Normal Distribution

- Normal distribution is a probability distribution where the probability of x is highest at the centre and lowest in the ends.

- Uniform distribution is a probability distribution where the probability of x is constant.

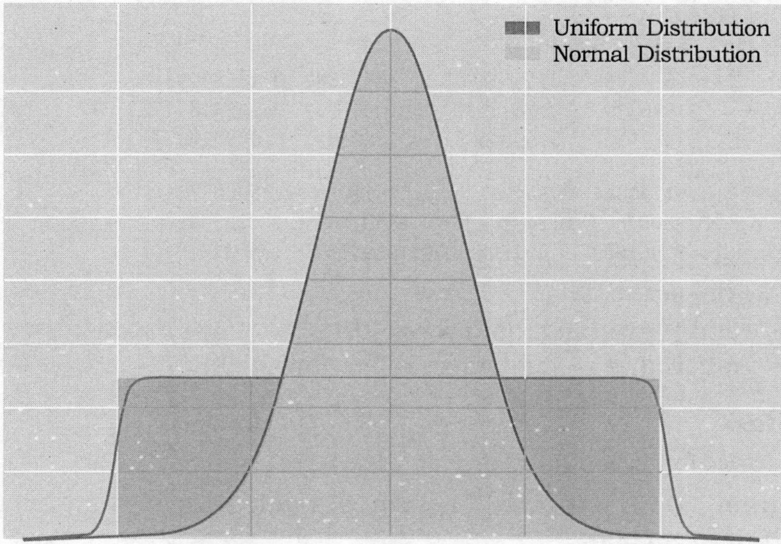

FIGURE 2.2   Uniform distribution vs normal distribution

(ii) rand(): It will generate random float values in the range [0,1) from uniform distribution samples. [0 ==> means 0 is included and 1) ==> means 1 is excluded.

Syntax

```
rand(d0, d1, ..., dn)
```

Random values in a given shape

Program 34: A program to print single random float value from 0 to 1.

```
import numpy as np9
print(np9.random.rand())
```

OUTPUT

```
C:\book>py h.py
0.9092916647042617
```

Program 35: A program to print 1-D array with float values from 0 to 1.

```
import numpy as np9
print(np9.random.rand(10))
```

OUTPUT

```
C:\book>py h.py
[0.51794159 0.98289473 0.99239635 0.09227589 0.49097997 0.27401264
 0.34612885 0.19441426 0.20043257 0.74575581]
```

Program 36: A program to print 3-D array with float values from 0 to 1.

```
import numpy as np9
print(np9.random.rand(2,3,4))
```

```
OUTPUT
C:\book>py  h.py
[[[0.65467552  0.55064775  0.86533882  0.18589418]
  [0.57643001  0.44481817  0.20946799  0.19941834]
  [0.94356274  0.69096082  0.10803837  0.03962945]]

 [[0.51740044  0.56496776  0.82511086  0.85008372]
  [0.58007627  0.86574438  0.93907441  0.8128393 ]
  [0.72561863  0.49156868  0.19590282  0.76622478]]]
```

(iii) uniform()

- rand() ==> range is always [0,1)
- uniform() ==> customize range [low,high)

Syntax

```
uniform(low=0.0,  high=1.0,  size=None)
```

Program 37: A program to print a random float number using uniform() function.

```python
import numpy as np9
print(np9.random.uniform()) # it will  acts  same  as  np9.random.
rand()
```

```
OUTPUT
C:\book>py  h.py
0.9404791837500721
```

Program 38: A program to print a random float value in the given customized range.

```python
import numpy as np9
print(np9.random.uniform(10,20))
```

```
OUTPUT
C:\book>py  h.py
15.816574043742694
```

Program 39: A program to print 1-D array with customized range of float values.

```python
import numpy as np9
print(np9.random.uniform(10,20,size=10))
```

```
OUTPUT
C:\book>py  h.py
[18.1310384  14.10395587  13.3380287  15.79094315  17.20007353
14.47280544
  19.88410912  11.42031482  16.94662155  11.44077793]
```

Program 40: A program to print 2-D array with customized range of float values.

```python
import numpy as np9
print(np9.random.uniform(10,20,size=(3,5)))
```

OUTPUT

```
C:\book>py  h.py
[[19.10672995 13.90164793 11.08141307 15.50829402 12.32057353]
 [18.50717084 18.41184221 17.9648871  13.86784327 15.34135452]
 [12.6092356  10.95898169 19.6283303  12.48246095 10.40802646]]
```

Program 41: A program to print 3-D array with customized range of float values.

```
import numpy as np9
print(np9.random.uniform(10,20,size=(2,3,4)))
```

OUTPUT

```
C:\book>py  h.py
[[[15.39664934 16.22045035 13.4218472  13.96214956]
  [10.4594827  19.36971362 17.53866178 17.99275043]
  [19.64628201 19.83391532 12.11000395 19.37707888]]

 [[19.7140768  17.39906017 16.89100872 14.17353024]
  [18.64570859 16.93985544 19.2405709  16.55389174]
  [17.71634179 11.37734884 10.03847467 18.96663334]]]
```

Program 42: A program to find the unifrom distribution in data visualization.

```
import numpy as np9
import matplotlib.pyplot as plt
s = np9.random.uniform(20,30,size=1000000)
count, bins, ignored = plt.hist(s, 15, density=True)
plt.plot(bins, np9.ones_like(bins), linewidth=2, color='r')
plt.show()
```

OUTPUT

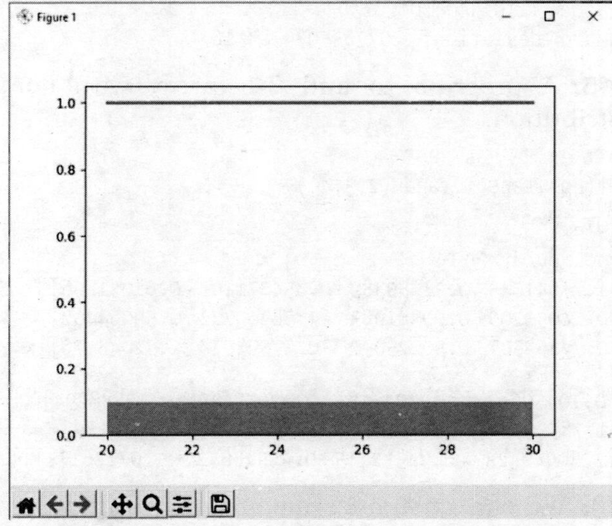

FIGURE 2.3 Unifrom distribution in data visualization

(iv) randn(): The values from normal distribution with mean = 0 and variance (standard deviation) = 1.

Syntax

```
randn(d0, d1, ..., dn)
```

Program 43: A program to find a single float value including –ve value.

```
import numpy as np9
a = np9.random.randn()
print(a)
```

OUTPUT

```
C:\book>py h.py
0.7501336304615012
```

Program 44: A program to find 1-D array with mean=0 and standard deviation=1.

```
import numpy as np9
a = np9.random.randn(10)
print(a)
print("Mean ", a.mean())
print("Variance ",a.var())
print("Standard deviation ",a.std())
```

OUTPUT

```
C:\book>py h.py
[-0.09698142  0.83851884  -0.68160955  -1.66820751  1.50901544
0.96881446
   0.72653551 -1.02128131 1.87966362 0.33067593]
Mean 0.27851440240634173
Variance 1.1613404701202303
Standard deviation 1.077655079383116
```

Program 45: A program to find 3-D array with float values in normal distribution.

```
import numpy as np9
print(np9.random.randn(2,3,4))
```

OUTPUT

```
C:\book>py h.py
[[[-1.77641073 0.31689389 0.85477116 -0.61051896]
  [-0.14030209 0.00381983 -1.56200772 0.59064402]
  [-0.77169907 0.50225097 -0.57593578 1.92048376]]

 [[ 0.10450667 -1.92045184 0.32083136  0.59888994]
  [-1.783262    1.46810093 1.32851619 -1.1961898 ]
  [ 2.50940423  0.94332215 0.95948642 -1.07276153]]]
```

(v) normal(): We can customize mean and variance.

Syntax

```
normal(loc=0.0, scale=1.0, size=None)
• loc : float or array_like of floats
Mean ("centre") of the distribution.
• scale : float or array_like of floats
Standard deviation (spread or "width") of the distribution. Must
be non-negative.
• size: int or tuple of ints, optional
```

Program 46: A program to print a random float number using normal() function.

```
import numpy as np9
print(np9.random.normal())
```

OUTPUT

```
C:\book>py h.py
-0.7256715761343828
```

Program 47: A program to print 1-D array with float values in normal distribution.

```
import numpy as np9
a = np9.random.normal(10,4,size=10)
print(" a ", a)
print("Mean ",a.mean())
print("Variance ",a.var())
print("Standard deviation ", a.std())
```

OUTPUT

```
C:\book>py h.py
a [10.73613946 -0.20776668 12.01777507 13.29751606 6.08469343
2.98295065
 6.95425722 2.64669252 10.14939879 15.70575111]
Mean 8.036740764040692
Variance 24.163576488828625
Standard deviation 4.915646090681125
```

Program 48: A program to print 2-D array with float values in normal distribution.

```
import numpy as np9
print(np9.random.normal(10,4,size=10))
```

OUTPUT

```
C:\book>py h.py
[ 2.44807269 8.28652409 6.82105486 9.90574003 11.37453896
 11.93105072 9.72750625 5.52921907 9.61494304 10.19849681]
```

Program 49: A program to print 3-D array with float values in normal distribution.

```
import numpy as np9
print(np9.random.normal(10,4,size=(2,3,4)))
```

OUTPUT

```
C:\book>py h.py
[[[ 3.03526162  6.68622315  10.79611049  18.45767697]
  [ 1.66939079  2.99384841   8.25740478   8.97832658]
  [13.14894766  2.79636781   1.47079875   9.17401125]]

 [[10.84455778  11.97473027  15.0167947   6.50166823]
  [10.48396039   9.81782556  21.19950873  8.84653567]
  [ 8.21366239  12.62796224   7.07489106  5.29512459]]]
```

Program 50: A program to find the uniform distribution in data visualization.

```python
import numpy as np9
import matplotlib.pyplot as plt
s = np9.random.normal(10,4,1000000)
count, bins, ignored = plt.hist(s, 15, density=True)
plt.plot(bins, np9.ones_like(bins), linewidth=2, color='r')
plt.show()
```

OUTPUT

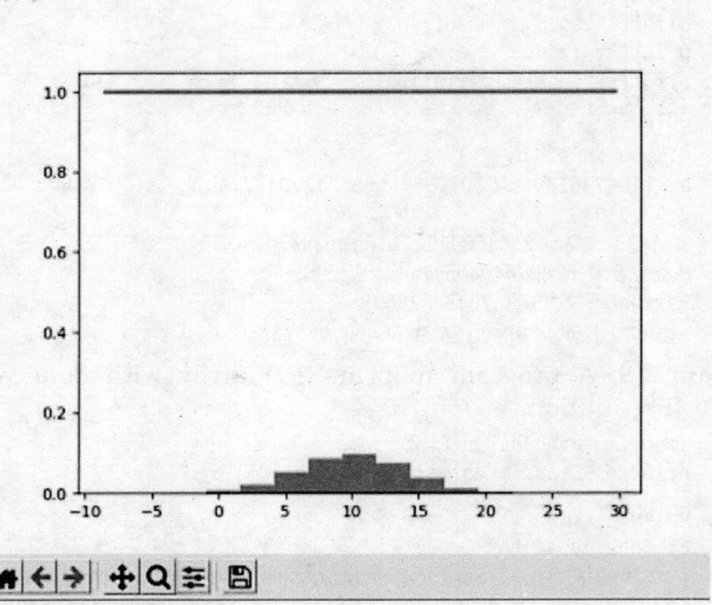

FIGURE 2.4   Output of uniform distribution

(vi) shuffle(): It modifies a sequence by shuffling its contents. This function only shuffles the array along the first axis of a multi-dimensional array (axis-0). The order of sub-arrays is changed but their contents remain the same.

Syntax

```
shuffle(...)
    method of numpy.random.mtrand.RandomState instance shuffle(x)
```

Program 51: A program to print 1-D array using shuffle() function.

```
import numpy as np9
a = np9.arange(9)
print('a before shuffling ',a)
np9.random.shuffle(a)
print('a after shuffling ',a)
```

OUTPUT

```
C:\book>py h.py
a before shuffling [0 1 2 3 4 5 6 7 8]
a after shuffling [8 5 4 6 1 2 3 0 7]
```

Program 52: A program to print 2-D array using shuffle() function.

```
import numpy as np9
a = np9.random.randint(1,101,size=(6,5))
print('a before shuffling ', a)
np9.random.shuffle(a) # inline shuffling happens
print('a after shuffling ', a)
```

OUTPUT

```
C:\book>py h.py
a before shuffling [[30 72 87 55  4]
 [10 75 53  2 88]
 [37 56 70 11 13]
 [ 8 73  2 21 61]
 [14 12 84 39 29]
 [29 13 37  1 77]]
a after shuffling [[14 12 84 39 29]
 [29 13 37  1 77]
 [10 75 53  2 88]
 [30 72 87 55  4]
 [37 56 70 11 13]
 [ 8 73  2 21 61]]
```

Program 53: A program to print 3-D array using shuffle() function.

```
#3-D array : (4,3,4)
• 4 : number of 2-D arrays (axis-0) ==> Vertical
• 3 : rows in every 2-D array(axis-1) ==> Horizontal
• 4 : columns in every 2-D array(axis-1)

import numpy as np9
a = np9.arrange(48).reshape(4,3,4)
print ('Before shuffling 3-D array', a)
np9.random.shuffle(a)
print('After shuffling 3-D array', a)
```

If we apply shuffle for 3-D array, then the order of 2-D arrays will be changed but the internal content remains the same.

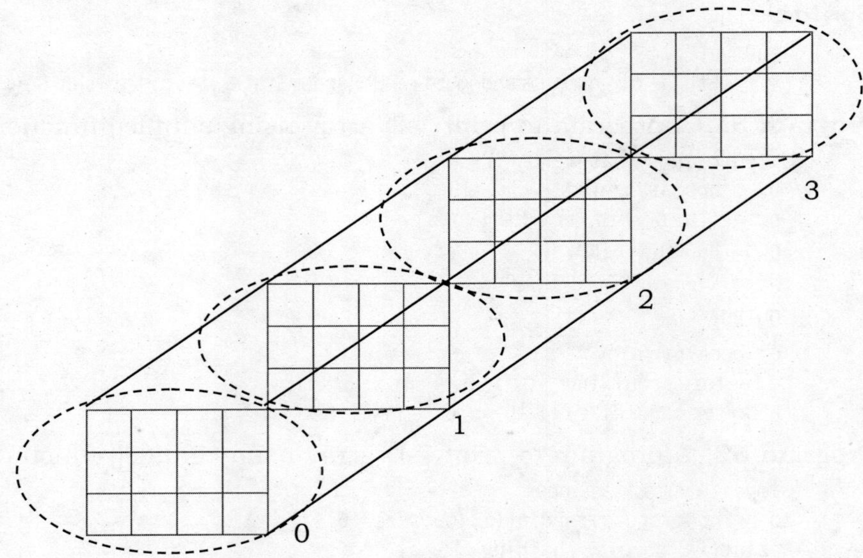

FIGURE 2.5   3-D array with shuffle()

OUTPUT

```
C:\book>py h.py
Before shuffling 3-D array [[[ 0  1  2  3]
  [ 4  5  6  7]
  [ 8  9 10 11]]

 [[12 13 14 15]
  [16 17 18 19]
  [20 21 22 23]]

 [[24 25 26 27]
  [28 29 30 31]
  [32 33 34 35]]

 [[36 37 38 39]
  [40 41 42 43]
  [44 45 46 47]]]
After  shuffling 3-D array [[[ 0  1  2  3]
  [ 4  5  6  7]
  [ 8  9 10 11]]

 [[24 25 26 27]
  [28 29 30 31]
  [32 33 34 35]]

 [[36 37 38 39]
  [40 41 42 43]
  [44 45 46 47]]
```

```
[[12 13 14 15]
 [16 17 18 19]
 [20 21 22 23]]]
```

Summary of Random Library Functions

1. randint() ==> To generate random int values in the given range
2. rand() ==> To generate uniform distributed float values in [0,1)
3. uniform() ==> To generate uniform distributed float values in the given range [low,high)
4. randn() ==> Normal distributed float values with mean=0 and standard deviation=1.
5. normal() ==> Normal distributed float values with specified mean and standard deviation.
6. shuffle() ==> To shuffle order of elements in the given nd array.

2.3 CREATING AND CONVERTING ARRAYS OF NUMPY DATA TYPES

NumPy has extra data types in addition to Python data types like int, float, complex, bool, str, etc. We can represent data types by using a single character also.

Table 2.3 NumPy data types

i	integer (int8, int16, int32, int64)
b	Boolean
u	unsigned integer (uint8, uint16, uint32, uint64)
f	float (float16, float32, float64)
c	complex (complex64, complex128)
s	String
U	Unicode string 51

int8

The value will be represented by using 8 bits.
The range of values is from −128 to 127.

int16

The value will be represented by using 16 bits.
The range of values is from −32768 to 32767.

int32

The value will be represented by using 32 bits.
The range of values is from –2147483648 to 2147483647.

int64

The value will be represented by using 64 bits.
The range of values is from –9223372036854775808 to 9223372036854775807.

Note:
    1. By default int means int32
    2. By default float means float64
    3. int8 type array requires less memory than int32 type array.

Example

```
>>> a = np9.array([10,20,30,40])
>>> import sys
>>> sys.getsizeof(a)
120
>>> a = np9.array([10,20,30,40],dtype='int8')
>>> sys.getsizeof(a)
108
```

How to check the data type of an array?

By using dtype attribute

Example

```
>>> a = np9.array([10,20,30,40])
>>> a.dtype
dtype('int32')
>>> a = np9.array([10.5,20.6,30])
>>> a.dtype
dtype('float64')
```

How to create array with required data type?

By using dtype argument

Example

```
>>> a = np9.array([10,20,30,40],dtype='i8')
>>> a.dtype
dtype('int64')
>>> a = np9.array([10,20,30,40],dtype='int8')
>>> a.dtype
dtype('int8')
>>> a = np9.array([10,20,30,40],dtype='i')
>>> a.dtype
dtype('int32')
```

Note: If the array element is unable to convert into the specified type, then we will get error.

Example

```
>>> a = np9.array(['a',10,10.5],dtype=int)
Traceback (most recent call last):
File "<stdin>", line 1, in <module>
ValueError: invalid literal for int() with base 10: 'a'
```

How to convert the type of existing array?

In general, we can use astype() function to convert the existing data type to another data type.

Example

```
>>> a = np9.array([10,20,30])
>>> a.dtype
dtype('int32')
>>> b = a.astype('float64')
>>> b
array([10., 20., 30.])
>>> b.dtype
dtype('float64')

>>> a = np9.array([10,20,30])
>>> a.dtype
dtype('int32')
>>> c = np9.float64(a)
>>> c
array([10., 20., 30.])
>>> c.dtype
dtype('float64')

>>> a = np9.array([10,20,30,0,40])
>>> a.dtype
dtype('int32')
>>> x = np9.bool_(a)
>>> x
array([ True, True, True, False, True])
>>> x.dtype
dtype('bool')
```

Program 54: A program to convert data type into int data type.

```
import numpy as np9
a = np9.array([10,20,30.5],dtype=int)
print(a)
```

OUTPUT

```
C:\book>py h.py
[10 20 30]
```

Program 55: A program to convert int data type into float data type.

```
import numpy as np9
a = np9.array([10,20,30.5],dtype=float)
print(a)
```

OUTPUT

```
C:\book>py h.py
[10. 20. 30.5]
```

Program 56: A program to convert the value to Boolean data type.

```
import numpy as np9
a = np9.array([10,20,30.5],dtype=bool)
print(a)
```

OUTPUT

```
C:\book>py h.py
[ True True True]
```

Program 57: A program to convert the value to complex type.

```
import numpy as np9
a = np9.array([10,20,30.5,0],dtype=complex)
print(a)
```

OUTPUT

```
C:\book>py h.py
[10. +0.j 20. +0.j 30.5+0.j 0. +0.j]
```

Program 58: A program to convert the value to str type.

```
import numpy as np9
a = np9.array([10,20,30.5],dtype=str)
print(a)
```

OUTPUT

```
C:\book>py h.py
['10' '20' '30.5']
```

2.4   CREATING OBJECT TYPE ARRAY

An object is the parent of all int, float, bool, complex, and str. Here the elements look heterogeneous. But the data type of elements is 'object' .

Program 59: A program to convert the value into object type.

```
import numpy as np9
a = np9.array([10,'Hima',10.5,True,10+20j],dtype=object)
print(a)
print("Data type of elements of array a ",a.dtype)
```

OUTPUT

```
C:\book>py h.py
[10 'Hima' 10.5 True (10+20j)]
Data type of elements of array a object
```

Creating arrays with specific data type

We have to use dtype argument while creating the array.

Program 60: A program to find the type of data type.

```
import numpy as np9
a = np9.random.randint(1,11,size=(20,30))
print("Data type of the elements ",a.dtype)
```

OUTPUT

```
C:\book>py h.py
Data type of the elements int32
```

astype()

We can convert the data type to another data type. By default, randint() will return int data type. By using "astype()", we can convert one data type to another.

Program 61: A program to convert int data type into float data type.

```
import numpy as np9
a = np9.random.randint(1,11,size=(20,30))
print("Before Conversion ")
print("Data type of a ",a.dtype)
b = a.astype('float')
print("Ater calling astype() on the ndarray a ")
print("Data type of b",b.dtype)
```

OUTPUT

```
C:\book>py h.py
Before Conversion
Data type of a int32
Ater calling astype() on the ndarray a
Data type of b float64
```

2.5   ACCESSING ELEMENTS OF NUMPY ARRAY

There are following two ways to access elements of NumPy array:

1. Indexing
2. Slicing

2.5.1   Indexing

By using an index, we can access a single element of the array. NumPy follows zero-based indexing, i.e., index of the first element is always zero. NumPy supports both positive indexing and negative indexing.

Syntax

```
a[index] --->Returns element present at specified index.
```

Example

```
>>> a = np9.array([10,20,30,40,50])
>>> a
array([10, 20, 30, 40, 50])
>>> a[0]
10
>>> a[2]
30
>>> a[-1]
50
>>> a[-2]
40
```

Note: If we are trying to access an array element with out-of-range index, then we will get IndexError.

Example

```
>>> a[10]
Traceback (most recent call last):
File "<stdin>", line 1, in <module>
IndexError: index 10 is out of bounds for axis 0 with size 5
>>> a[-7]
Traceback (most recent call last):
File "<stdin>", line 1, in <module>
IndexError: index -7 is out of bounds for axis 0 with size 5
```

Accessing elements of 2-D array

Syntax

```
a[i][j]
where i-->means row index
      j-->means column index
```

Example

```
>>> a = np9.array([[10,20,30],[40,50,60]])
>>> a
array([[10, 20, 30],
[40, 50, 60]])
```

Accessing elements of 3-D array

3-D array means array of 2-D arrays.

Syntax

```
a[i][j][k]
where i represents which 2-Dimensional array
      j represents row number in that 2-D array
      k represents column number in that 2-D array
```

Example
```
a[0][1][2]
0 index 2-D array
In that 2-D array, 1 index row number
In that 2-D array, 2-index column number
```

Example
```
>>> l = [[[1,2,3],[4,5,6],[7,8,9]],[[10,11,12],[13,14,15],[16,17,18]]]
>>> a = np9.array(l)
>>> a
array([[[  1,  2,  3],
        [  4,  5,  6],
        [  7,  8,  9]],
       [[10,  11,  12],
        [13,  14,  15],
        [16,  17,  18]]])
>>> a.shape
(2, 3, 3)
2 ---> Number of 2-D arrays
3 ---> in every 2-D arrays, 3 number of rows
3 ---> in every 2-D array, 3 number of columns
```

Example
```
l = [[[1,2,3,4],[5,6,7,8],[9,10,11,12]],[[13,14,15,16],[17,18,19,20],[21,
22,23,24]]]
a = np9.array(l)

2--->2-D arrays
3 --->3 Rows in every 2-D array
4 ---> 4 Columns in every 2-D Array
shape=(2,3,4)
```

It is 3-D array which contains two 2-D arrays. Each 2-D array contains 3 rows and 4 columns. Total 24 elements are there.

```
        a[i][j][k]
where
        i--->which 2D array
        j--->In that 2D array, Row index
        k --->In that 2D array, column index
```

Example
```
>>> l = [[[1,2,3,4],[5,6,7,8],[9,10,11,12]],[[13,14,15,16],[17,18,19,20],[21,22,23,24]]]
>>> a = np9.array(l)
>>> a
array([[[  1,  2,  3,  4],
        [  5,  6,  7,  8],
        [  9, 10, 11, 12]],
       [[13, 14, 15, 16],
        [17, 18, 19, 20],
        [21, 22, 23, 24]]])
```

Accessing elements of 4-D array

4-D array means an array of 3-D arrays.

Syntax

```
a[i][j][k][l]
```
where
```
i-->which 3d array
j-->in that 3d array, which 2d array
k--->in that 2d array which row
l --->in that 2d array which column
```

2.5.2 Slicing

Slice means part of the object.

Slice Operator on Python List

Example

```
>>> l=[10,20,30,40,50,60,70]
>>> l
[10, 20, 30, 40, 50, 60, 70]
```

Syntax-1

```
l[begin:end] --->It returns elements from begin index to end-1 index.
    The default value for begin: 0
    The default value for end: len(l)
```

Example

```
>>> l[1:6] ---->It returns elements from 1st index to 5th index
[20, 30, 40, 50, 60]
>>> l[-6:-2] ---->It returns elements from -6th index to -3 index
[20, 30, 40, 50]
>>> l[:4] ---->It returns elements from 0 index to 3 index
[10, 20, 30, 40]
>>> l[2:]--->It returns elements from 2nd index to last element
[30, 40, 50, 60, 70]
>>> l[4:1]
[]
>>> l[4:10000]
[50, 60, 70]
>>> l[-10:]
[10, 20, 30, 40, 50, 60, 70]
>>>l[:]
[10, 20, 30, 40, 50, 60, 70]
```

Note: Slice operator won't raise IndexError.

Syntax-2

```
l[begin:end:step]
```

where
        it returns elements from begin index to end-1 index based on step value.
        Step is optional and default value 1.

Example

l[1:6:1]--->It returns elements from 1st index to 5th index and every element will be considered.
l[1:6:2]--->It returns elements from 1st index to 5th index and every 2nd element will be considered.
l[1:6:3]--->It returns elements from 1st index to 5th index and every 3rd element will be considered.

```
>>> l[1:6:1]
[20, 30, 40, 50, 60]
>>> l[1:6:2]
[20, 40, 60]
>>> l[1:6:3]
[20, 50]
```

Rules of Slice Operator

1. All the three parameters of the slice operator are optional.
2. For begin, end, step attributes, we can take both positive and negative values. But for step, we cannot take zero.

```
>>> l[1:6:0]
```

Traceback (most recent call last):
    File "<stdin>", line 1, in <module>
    ValueError: slice step cannot be zero

3. Based on step value, we can decide whether we have to consider elements in either forward or backward direction.
   If the step value is positive, we have to consider elements from begin to end-1 in the forward direction.
   If the step value is negative, we have to consider elements from begin to end+1 in the backward direction.
4. In the forward direction, if the end value is 0 then the result is always empty.
5. In the backward direction, if the end value is -1 then the result is always empty.

Example

l[5:1:-1] ---> Return elements from 5th index to 2nd index in backward direction
l[5:1:-2] --->Return elements from 5th index to 2nd index in backward direction, every second element we have to consider.
l[4:-5:1]--->from 4th index to -6 in forward direction

```
>>> l[5:1:-1]
[60, 50, 40, 30]
>>>
```

```
>>> l[5:1:-2]
    [60, 40]
>>> l[4:-5:1]
[]
l[-2:-6:-1]---> from -2 to -5 in backward direction
l[5:1:-1]--->from 5th index to 2nd index in backward direction
l[-1::-2]--->from -1 index to beginning of the list, every 2nd element
in backward direction
l[::]--->All elements in forward direction
l[::-1] ---> All elements in backward direction
>>> l[-2:-6:-1]
    [60, 50, 40, 30]
>>> l[5:1:-1]
    [60, 50, 40, 30]
>>> l[-1::-2]
    [70, 50, 30, 10]
>>> l[::]
    [10, 20, 30, 40, 50, 60, 70]
>>> l[::-1]
    [70, 60, 50, 40, 30, 20, 10]
    l[-1:-1:-1]--->Empty
```

Slice Operator on 1-D Numpy Array

The rules are exactly same as Python's built-in slice operator.

Syntax

```
a[begin:end:step]
```

Rules

1. All the three parameters of the Slice operator are optional.
2. For begin, end, step attributes, we can take both positive and negative values.
   But for step we cannot take zero.
   ```
   >>> l[1:6:0]
   ```
   Traceback (most recent call last):
   ```
   File "<stdin>", line 1, in <module>
   ValueError: slice step cannot be zero
   ```
3. Based on step value, we can decide whether we have to consider elements in either the forward or backward direction. If the step value is positive, we have to consider elements from begin to end-1 in a forward direction. If the step value is negative, we have to consider elements from begin to end+1 in a backward direction.
4. In the forward direction, if the end value is 0 then the result is always empty.
5. In the backward direction, if the end value is –1 then the result is always empty.

Example

```
>>> a = np9.array([10,20,30,40,50,60,70])
>>> a
array([10, 20, 30, 40, 50, 60, 70])
Diagram
>>> a[2:5]
array([30, 40, 50])
>>> a[:5]
array([10, 20, 30, 40, 50])
>>> a[2:]
array([30, 40, 50, 60, 70])
>>> a[::2]
array([10, 30, 50, 70])
```

Slice Operator on 2-D Numpy Array

Example

```
>>> a = np9.array([[10,20],[30,40],[50,60]])
>>> a
array([[10, 20],
[30, 40],
[50, 60]])
```

Syntax

```
        arrayname[row,column]
        arrayname[begin:end:step,begin:end:step]
```

The first slice operator talks about rows and second slice operator talks about columns.

Example 1

```
>>> a[:,0:1]
array([[10],
[30],
[50]])
>>> a[:,:1]
array([[10],
[30],
[50]])
```

Note:

- For both row and column, we should use the slice operator only. If we use the index directly, we may get 1-D array as the result, but not 2-D array.

Program 62: A program to print p-nd array properties.

```
        import numpy as np9
        l=[60,70,80]
        a=np9.array(l)
        print(a.ndim)
```

```
print(a.dtype)
print(a.size)
print(a.shape)
```

OUTPUT

```
C:\book>py h.py
1
int32
3
(3,)
```

Program 63: A program to print 1-D array using tuples.

```
import numpy as np9
t = ('hima1','hima2','hima3')
print('Type of t ',type(t))
a = np9.array(t)
print("tyye of a",type(a))
print('The data type of elements of array a',a.dtype)
print("size...",a.size)
print("Shape...,",a.shape)
print(a)
```

OUTPUT

```
C:\book>py h.py
Type of t <class 'tuple'>
tyye of a <class 'numpy.ndarray'>
The data type of elements of array a <U5
size... 3
Shape..., (3,)
['hima1' 'hima2' 'hima3']
```

Program 64: A program to create 2-dimensional array for the given nested list.

```
[[10,20,30],[40,50,60],[70,80,90]]
import numpy as np9
a = np9.array([[10,20,30],[40,50,60],[70,80,90]])
print('Type of a ',type(a))
print('a ==> \n', a)
print('The dimensions of the array a ',a.ndim)
print('The data type of elements of array a',a.dtype)
print('The size of the array a ',a.size)
print('The shape of the array a ',a.shape)
```

OUTPUT

```
C:\book>py h.py
Type of a <class 'numpy.ndarray'>
a ==>
 [[10 20 30]
  [40 50 60]
  [70 80 90]]
The dimensions of the array a 2
The data type of elements of array a int32
```

```
The size of the array a 9
The shape of the array a (3, 3)
```

Note:

- Array contains only homogeneous elements.
- If a list contains heterogeneous elements, then while creating the array, upcasting will be performed.

Program 65: A program to find a list containing heterogeneous elements.

```
import numpy as np9
l = [10,20,30.5]
a = np9.array(l) # upcasting to float
print('a :: ',a)
print('data type of elements of a ...' ,a.dtype)
```

OUTPUT

```
C:\book>py h.py
a :: [10. 20. 30.5]
data type of elements of a ... float64
```

EXERCISES

1. Write a program for printing 1-D array using NumPy (start, end and step values: start to end-1 with step).

2. Write a program to print 4 evenly spaced values between 0 and 1 including 0 and excluding 1 and return spacing.

3. Write a program to print 5-D array using eye() function with +1 diagonal.

4. Write a program to construct +1 diagonal for given 2-D array by providing two arguments (first argument represents no. of elements and second element represents the diagonal) and remaining elements are filled with zeros.

5. Write a program to print 2-D array with float values 0 to 1.

6. Write a program to find 2-D array with float values in normal distribution.

7. Write a program to check whether the memory utilization is improved with dtype.

3

DATA TYPES AND STRING OPERATIONS

3.1 INTRODUCTION

The data types that are used in the NumPy library are as follows:

- **i** integer (int8, int16, int32, int64)
- **b** Boolean
- **u** unsigned integer (uint8, uint16, uint32, uint64)
- **f** float (float16, float32, float64)
- **c** complex (complex64, complex128)
- **s** string
- **U** unicode string
- **M** datetime and so on

The int data type is the default type of integer. The int data type will be represented by int8 (8 bits), int16 (16 bits), int32 (32 bits), and int64 (64 bits).

int8(i1)

The value of int8 will be represented by 8 bits. Most Significant Bit (MSB) is reserved for a sign. The value of int8 ranges from −128 to 127.

0	1	2	3	4	5	6	7
0	1	1	1	1	1	1	1

Represents Sign Bit   Max Value : 127
(Most Significant Bit (MSB))   Min Value : −128
0 : +ve   1 : −ve

FIGURE 3.1 Integer data type representation

int16(i2)

The value of int16 will be represented by 16 bits. Most Significant Bit (MSB) is reserved for a sign. The value of int16 ranges from –32768 to 32767.

int32 (i4)

The value of int32 will be represented by 32 bits. Most Significant Bit (MSB) is reserved for a sign. The value of int32 ranges from –2147483648 to 2147483647. It is the default data type.

int64

The value of int64 will be represented by 64 bits. Most Significant Bit (MSB) is reserved for a sign. The value of int64 ranges from –9223372036854775808 to 9223372036854775807.

Program 1: Write a program for the efficient usage of memory by considering the data types.

```
import sys
import numpy as np
a = np.array([110,120,130,140]) # default data type : int32
print(f"Size of int32 : {sys.getsizeof(a)}")
a = np.array([110,120,130,140],dtype='int8')
print(f"Size of int8 : {sys.getsizeof(a)}")
OUTPUT
C:\h_numpy>py h9.py
Size of int32 : 120
Size of int8 : 108
```

3.2   CHANGING THE DATA TYPE OF AN EXISTING ARRAY

We change the data type of an existing array in the following two ways:

- by using **astype()**
- by using built-in function of numpy like **float64()**

Program 2: Write a program to convert int to float data type by using astype().

```
import numpy as np
a = np.array([110,120,130,140])
b = a.astype('float64')
print(f"Datatype of elements of array a : {a.dtype}")
print(f"Datatype of elements of array b : {b.dtype}")
print(a)
print(b)

OUTPUT
C:\h_numpy>py h9.py
Datatype of elements of array a : int32
Datatype of elements of array b : float64
```

```
[110 120 130 140]
[110. 120. 130. 140.]
```

Program 3: Write a program to convert int to float data type by using built-in function float64().

```
import numpy as np
a = np.array([10,20,30,40])
b = np.float64(a)
print(f"Datatype of elements of array a : {a.dtype}")
print(f"Datatype of elements of array b : {b.dtype}")
print(a)
print(b)
```

OUTPUT

```
C:\h_numpy>py h9.py
Datatype of elements of array a : int32
Datatype of elements of array b : float64
[10 20 30 40]
[10. 20. 30. 40.]
```

Program 4: Write a program to convert int to Boolean data type by using built-in function bool.

```
import numpy as np
a = np.array([110,10,120,10,130])
x = np.bool(a)
```

OUTPUT

ValueError: The truth value of an array with more than one element is ambiguous. Use a.any() or a.all()

Another example

```
import numpy as np
a = np.array([110,10,120,0,130])
x = np.bool_(a)
print(x)
```

OUTPUT

```
C:\h_numpy>py h9.py
[ True True True False True]
```

3.3 VIEW

The view will not create a separate object and it is just "logical representation of existing array", i.e., it creates a reference to an existing array. If we perform any changes to the original array, those changes will be reflected in the original array and vice-versa. We can create a view explicitly by using the view() method of the nd array class.

Program 5: Write a program for changing the values of the original array using the view() method.

```
import numpy as np
a = np.array([10,20,30,40])
```

```
b = a.view()
print("Original Array(a)",a)
print("View of the Array(b) ",b)
print("*"*80)
print("Changing the values of a, it will reflect to view(b) also :")
a[0]=7777
print("Changed Array(a) ",a)
print("View of the Array(b) ",b)
print("*"*80)
print("Changing the values of view(b), it will reflect to a also :")
b[-1]=999
print("Changed view Array(b) ",b)
print("Original Array(a) ",a)
```

OUTPUT

```
C:\h_numpy>py h9.py
Original Array(a) [10 20 30 40]
View of the Array(b) [10 20 30 40]
********************************************************************
Changing the values of a, it will reflect to view(b) also :
Changed Array(a) [7777 20 30 40]
View of the Array(b) [7777 20 30 40]
********************************************************************
Changing the values of view(b), it will reflect to a also :
Changed view Array(b) [7777 20 30 999]
Original Array(a) [7777 20 30 999]
```

3.4 COPY

It will create an external copy of the data object. If we perform any changes to the original array, those changes won't be reflected to the copy and vice-versa. By using the copy() method of the nd array class, we can create a copy of the existing nd array.

3.5 NUMPY STRING OPERATIONS

This module is used to perform vectorized string operations for arrays of dtype np9.string_ or np9.unicode_. All of them are based on the standard string functions in Python's built-in library.

(i) numpy.lower()

This function returns the lowercase of the given string. It converts all uppercase characters to lowercase. If no uppercase characters exist then it returns the original string.

Example

```
import numpy as np9
print(np9.char.lower(['HIMA', 'EDUCATION']))
print(np9.char.lower('HIMA'))
```

```
OUTPUT
C:\h_np9>py h9.py
['hima' 'education']
Hima
```

(ii) numpy.split()

This function returns a list of strings after breaking the given string by the specified separator.

Example

```
import numpy as np9
print(np.char.split('The book of np9 by hima'))
print(np.char.split('The book, of np9, by hima', sep = ','))
```

```
OUTPUT
C:\h_np9>py h9.py
['The', 'book', 'of', 'np9', 'by', 'hima']
['The book', ' of np9', ' by hima']
```

(iii) numpy.join()

This function is a string method and returns a string in which the elements of a sequence have been joined by a str separator.

Example

```
import numpy as np9
print(np.char.join('-', 'hima'))
print(np.char.join(['-', ':'], ['hima', 'education']))
```

```
OUTPUT
C:\h_np9>py h9.py
h-i-m-a
['h-i-m-a' 'e:d:u:c:a:t:i:o:n']
```

(iv) numpy.count()

This function returns the number of occurrences of a substring in the given string.

Example

```
import numpy as np9
a=np9.array(['np9 ', 'by', 'hima'])
# counting a substring
print(np9.char.count(a,'hima'))
# counting a substring
print(np9.char.count(a, 'hi'))
```

```
OUTPUT
C:\h_np9>py h9.py
[0 0 1]
[0 0 1]
```

(v) numpy.rfind()

This function returns the highest index of the substring found in a given string. If it is not found, then it returns −1.

Example

```
import numpy as np9
a=np9.array(['np9', 'by', 'hima'])
# counting a substring
print(np9.char.rfind(a,'hima'))
# counting a substring
print(np9.char.rfind(a, 'hi'))
```

OUTPUT

```
C:\h_np9>py h9.py
[-1 -1 0]
[-1 -1 0]
```

(vi) numpy.isnumeric()

This function returns "True" if all characters in the string are numeric characters otherwise it returns "False".

Example

```
import numpy as np9
# counting a substring
print(np.char.isnumeric('hima'))
# counting a substring
print(np.char.isnumeric('9hima'))
```

OUTPUT

```
C:\h_np9>py h9.py
False
False
```

(vii) numpy.equal()

This function checks for string1 == string2 elementwise.

Example

```
import numpy as np9
# comparing a string elementwise
# using equal() method
a=np.char.equal('hima','Hima')
print(a)
```

OUTPUT

```
C:\h_np9>py h9.py
False
```

(vii) numpy.greater()

This function checks whether string1 is greater than string2 or not.

Example

```
import numpy as np9
# comparing a string elementwise
```

```
# using greater() method
a=np.char.greater('hima','np9')
print(a)
```

OUTPUT

```
C:\h_np9>py h9.py
False
```

Table 3.1 Other string operations

Function	Description
np9.find()	It returns the lowest index of the substring if it is found in given string. If it is not found then it returns −1.
np9.index()	It returns the position of the first occurrence of substring in a string.
np9.isalpha()	It returns "True" if all characters in the string are alphabets. Otherwise, it returns "False".
np9.isdecimal()	It returns "True" if all characters in a string are decimal. If all characters are not decimal then it returns "False".
np9.isdigit()	It returns "True" if all characters in the string are digits. Otherwise, it returns "False".
np9.islower()	It returns "True" if all characters in the string are lowercase. Otherwise, it returns "False".
np9.isspace()	It returns "True" for each element if there are only whitespace characters in the string and there is at least one character, otherwise returns "False".
np9.istitle()	It returns "True" for each element if the element is a titlecased string and there is at least one character, otherwise returns "False".
np9.isupper()	It returns "True" for each element if all cased characters in the string are uppercase and there is at least one character, otherwise returns "False".
np9.rindex()	It returns the highest index of the substring inside the string if substring is found. Otherwise it raises an exception.
np9.startswith()	It returns "True" if a string starts with the given prefix, otherwise returns "False".
np9.strip()	It is used to remove all the leading and trailing spaces from a string.
np9.capitalize()	It converts the first character of a string to capital (uppercase) letter. If the string has its first character as capital, then it returns the original string.
np9.center()	It creates and returns a new string which is padded with the specified character.

np9.decode()	It is used to convert encoding scheme, in which argument string is encoded to the desired encoding scheme.
np9.encode()	It returns the string in the encoded form.
np9.ljust()	It returns an array with the left-justified elements in a string of length/width.
np9.rjust()	For each element, it returns a copy with the leading characters removed.
np9.strip()	For each element, it returns a copy with the leading and trailing characters removed.
np9.lstrip()	It converts angles from degrees to radians.
np9.rstrip()	For each element, it returns a copy with the trailing characters removed.
np9.partition()	It splits each element around sep.
np9.rpartition()	It splits each element around the right-most separator.
np9.rsplit()	For each element, it returns a list of the words in the string, using sep as the delimiter string.
np9.title()	It is used to convert the first character in each word to uppercase and remaining characters to lowercase in string and returns new string.
np9.upper()	It returns the uppercased string from the given string. It converts all lowercase characters to uppercase. If no lowercase characters exist, it returns the original string.

EXERCISES

1. Write a program to copy the values of original array to another object using copy() method.

2. Write a program to capitalize the first letter, lowercase, uppercase, swapcase, title-case of all the elements of a given array.

3. Write a program to check whether each element of a given array starts with given alphabet or not.

4 INDEXING, SLICING AND ADVANCED INDEXING

4.1 ACCESSING ELEMENTS OF nd ARRAY

An array element can be accessed by the following methods:

- Indexing
- Slicing
- Advanced indexing
- Condition-based selection

Indexing—Only One Element

By using an index, we can access a single element of the array. In the zero-based indexing, the index of the first element is 0. It supports both +ve and –ve indexing.

Syntax

1-D array

```
array[index]
```

2-D array

```
array[row_index][column_index]
```

3-D array

```
array[2-D array index][row_index_in that 2-D array]
                                    [columnindex in that 2-d array]
```

Program 1: A program to print the value from 1-D array based on a given index value.

```
import numpy as np9
a = np9.array([10,20,30,40,50])
print(a[3])
```

OUTPUT

```
C:\h_numpy>py h9.py
40
```

Program 2: A program to print the value from 1-D array based on a given negative index value.

```
import numpy as np9
a = np9.array([10,20,30,40,50])
print(a[-1])
```

OUTPUT

```
C:\h_numpy>py h9.py
50
```

Accessing elements of 2-D array

A 2-D array is the collection of 1-D arrays. The two axes are present in the 2-D array. Axis-0 indicates row index and axis-1 indicates column index.

Synatx

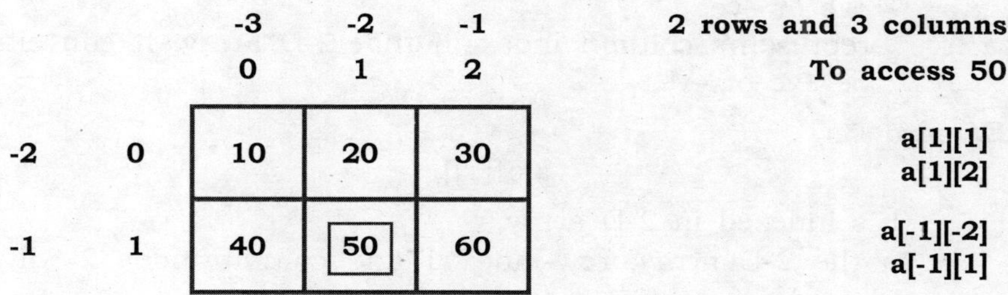

FIGURE 4.1   Accessing elements of 2-D array

Program 3: A program for accessing the elements of 2-D array.

```
import numpy as np9
a = np9.array([[110,120,130],[140,150,160]])
print("Shape of the array a : ",a.shape)
print("To Access the element 50")
print("a[1][1] ",a[1][1])
print("a[1][-2] ",a[1][-2])
print("a[-1][2] ",a[-1][-2])
print("a[-1][1] ",a[-1][1])
```

OUTPUT

```
C:\h_numpy>py h9.py
Shape of the array a : (2, 3)
To Access the element 50
a[1][1] 150
a[1][-2] 150
a[-1][2] 150
a[-1][1] 150
```

Accessing elements of 3-D array

A 3-D array is the collection of 2-D array. There are three indexes in 3-D array:

axis-0	No. of 2-D arrays
axis-1	Row index
axis-2	Column index

Syntax

```
array[2-D array index][row_index_in that 2-D array]
                           [colindex in that 2-d array]
```

Example

```
a[i][j][k]
```

where

i   represents 2-D array (index of 2-D array). It can either be +ve or −ve.

j   represents row index in that 2-D array. It can either be +ve or −ve.

k   represents column index in that 2-D array. It can either be +ve or −ve.

Example

```
a[0][1][2]
```

- 0 is indexed in 2-D array.
- In the 2-D array, row-index 1 and column-index 2 will be selected.

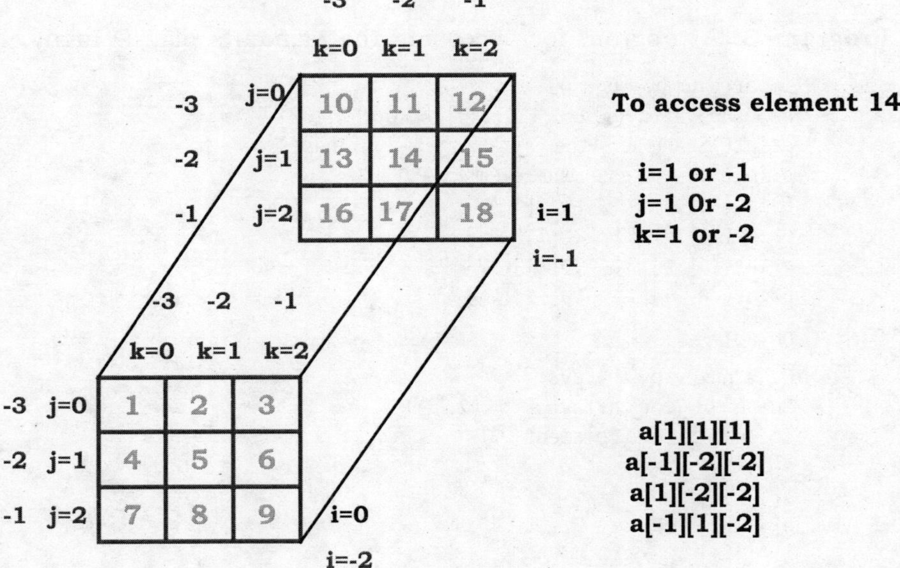

FIGURE 4.2 Accessing elements of 3-D array

Program 4: A program for accessing the elements of the 3-D array.

```
import numpy as np9
l = [[[1,2,3],[4,5,6],[7,8,9]],[[10,11,12],[13,14,15],[16,17,18]]]
a = np9.array(l)
print("Shape of the array a",a.shape)
print("To access the element 14 from the 3-D array")
print("a[1][1][1] ",a[1][1][1])
print("a[-1][-2][-2] ",a[-1][-2][-2])
print("a[1][-2][-2] ",a[1][-2][-2])
print("a[-1][1][-2] ",a[-1][1][-2])
```

OUTPUT

```
C:\h_numpy>py h9.py
Shape of the array a (2, 3, 3)
To access the element 14 from the 3-D array
a[1][1][1] 14
a[-1][-2][-2] 14
a[1][-2][-2] 14
a[-1][1][-2] 14
```

Accessing elements of 4-D array

A 4-D array contains multiple 3-D arrays. Every 3-D array contains multiple 2-D arrays. Every 2-D array contains rows and columns.

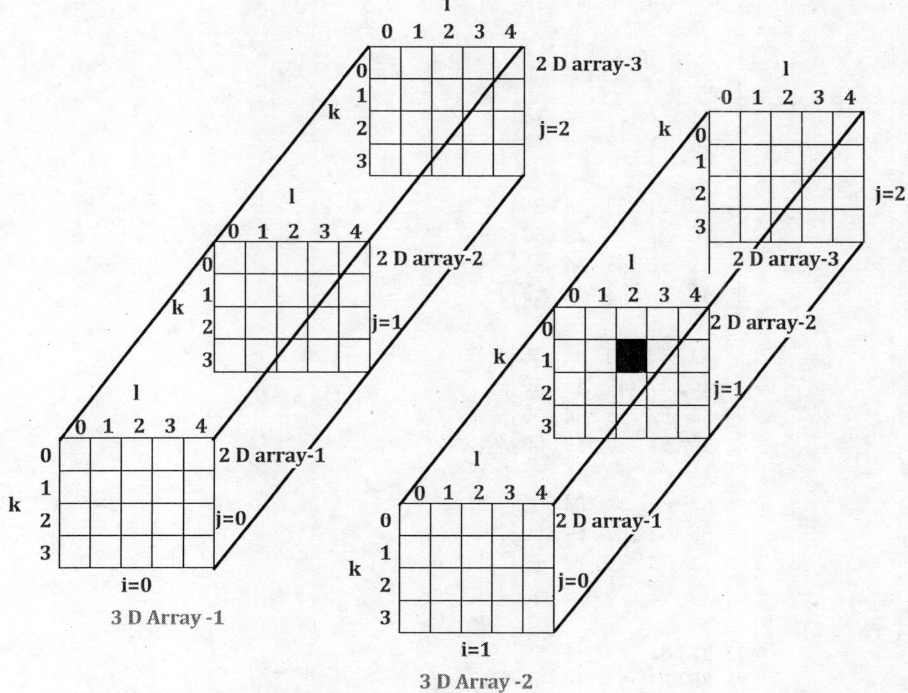

FIGURE 4.3   Accessing elements of 4-D array

Example

$$(i,j,k,l) \implies (2,3,4,5)$$

2   represents the number of 3-D arrays

3   every 3-D array contains three 2-D arrays

Every 2-D array contains 4 rows and 5 columns

$$a[i][j][k][l]$$

i   represents 3-D array

j   represents 2-D array in that 3-D array

k   row index of the 2-D array

l   column index of the 2-D array

Program 5: A program for accessing the elements of the 4-D array.

```
import numpy as np9
a = np9.arange(1,121).reshape(2,3,4,5)
print("Size of the array ",a.size)
print("Shape of the array ",a.shape)
print("a ==>\n ",a)
```

OUTPUT

```
C:\h_numpy>py h9.py
Size of the array 120
Shape of the array (2, 3, 4, 5)
a ==>
 [[[[  1   2   3   4   5]
   [  6   7   8   9  10]
   [ 11  12  13  14  15]
   [ 16  17  18  19  20]]

  [[ 21  22  23  24  25]
   [ 26  27  28  29  30]
   [ 31  32  33  34  35]
   [ 36  37  38  39  40]]

  [[ 41  42  43  44  45]
   [ 46  47  48  49  50]
   [ 51  52  53  54  55]
   [ 56  57  58  59  60]]]

 [[[ 61  62  63  64  65]
   [ 66  67  68  69  70]
   [ 71  72  73  74  75]
   [ 76  77  78  79  80]]

  [[ 81  82  83  84  85]
   [ 86  87  88  89  90]
   [ 91  92  93  94  95]
   [ 96  97  98  99 100]]
```

```
[[101  102  103  104  105]
 [106  107  108  109  110]
 [111  112  113  114  115]
 [116  117  118  119  120]]]]
```

Program 6: A program to print the accessing elements at a[1][1][1][2].

```
import numpy as np9
a = np9.arange(1,121).reshape(2,3,4,5)
print("The element present at a[1][1][1][2] ",a[1][1][1][2])
```

OUTPUT

```
C:\h_numpy>py h9.py
The element present at a[1][1][1][2] 88
```

4.2 SLICING

Slicing means the grouping of elements that are in order.

Python's Slice Operator

$$l[begin:end]$$

It returns elements from begin index to (end-1) index.

-7   -6   -5   -4   -3   -2   -1

[10, 20, 30, 40, 50, 60, 70]

0   1   2   3   4   5   6

FIGURE 4.4   Indexing values in slicing

Example

```
l= [10,20,30,40,50,60,70]
print(l[2:6]) #from 2nd index to 5th index
```

OUTPUT

```
C:\h_numpy>py h9.py
[30, 40, 50, 60]
```

Example

```
l= [10,20,30,40,50,60,70]
print(l[-6:-2]) #From -6 index to -3 index
```

OUTPUT

```
C:\h_numpy>py h9.py
[20, 30, 40, 50]
```

Example

```
# If we are not specifying begin index then default is 0
l= [10,20,30,40,50,60,70]
print(l[:3]) #from 0 index to 2nd index
```

OUTPUT

```
C:\h_numpy>py h9.py
[10, 20, 30]
```

Example

```
# If we are not specifying end index then default is upto last
l= [10,20,30,40,50,60,70]
print(l[2:]) # from 2nd index to last
```

OUTPUT

```
C:\h_numpy>py h9.py
[30, 40, 50, 60, 70]
```

Example

```
# if we are not specifying begin and end then it will considered
all elements
l= [10,20,30,40,50,60,70]
print(l[:]) # it is equivalent to l
```

OUTPUT

```
C:\h_numpy>py h9.py
[10, 20, 30, 40, 50, 60, 70]
```

Syntax-2

```
l[begin:end:step]
```

where

- begin, end and step values can either be +ve or –ve. These are optional.
- If begin is not specified then 0 will be taken by default.
- If the end is not specified then up to the end of the list will be taken.
- If step is not specified then 1 will be taken by default.
- The step value cannot be 0.
- If step value is +ve, then begin to end–1 in forwarding direction will be considered.
- If step value is –ve, then begin to end+1 in backward direction will be considered.
- In the forward direction, if the end is 0 then the result will be always empty.
- In the backward direction, if the end is –1 then the result will be always empty.

Note:

- The slice operator does not raise any **IndexError** when the index is out of range.

Example:

```
# slice operator in python list
l = [10,20,30,40,50,60,70,80,90]
print(l[2:6:2]) # from 2nd index to 6-1 index with step 2
```

OUTPUT

```
C:\h_numpy>py h9.py
[30, 50]
```

Slice operator on 1-D array

It follows the same rules like in Python slice operator.

Syntax

```
array_name[begin:end:step]
```

where

- begin, end and step values can either be +ve or –ve. These are optional.
- If begin is not specified then 0 will be taken by default.
- If the end is not specified then up to the end of the list will be taken.
- If step is not specified then 1 will be taken by default.
- The step value cannot be 0.
- If step value is +ve, then begin to end–1 in forwarding direction will be considered.
- If step value is –ve, then begin to end+1 in backward direction will be considered.
- In the forward direction, if the end is 0 then the result will be always empty.
- In the backward direction, if the end is –1 then the result will be always empty.

Program 7: A program to print values using Slice operator on 1-D array.

```
import numpy as np9
a = np9.arange(10,101,10)
print(a)
```

OUTPUT

```
C:\h_numpy>py h9.py
[  10  20  30  40  50  60  70  80  90  100]
```

Program 8: A program to print values using Slice operator from 2nd index to 5-1 index.

```
import numpy as np9
a = np9.arange(10,101,10)
print(a[2:5] )
```

OUTPUT

```
C:\h_numpy>py h9.py
[30  40  50]
```

Program 9: A program to print values using Slice operator on an array.

```
import numpy as np9
a = np9.arange(10,101,10)
print(a[::1] )
```

OUTPUT

```
C:\h_numpy>py h9.py
[  10  20  30  40  50  60  70  80  90  100]
```

Slice operator on 2-D NumPy array

Syntax

- array_name[row,column]
- array_name[begin:end:step, begin:end:step]

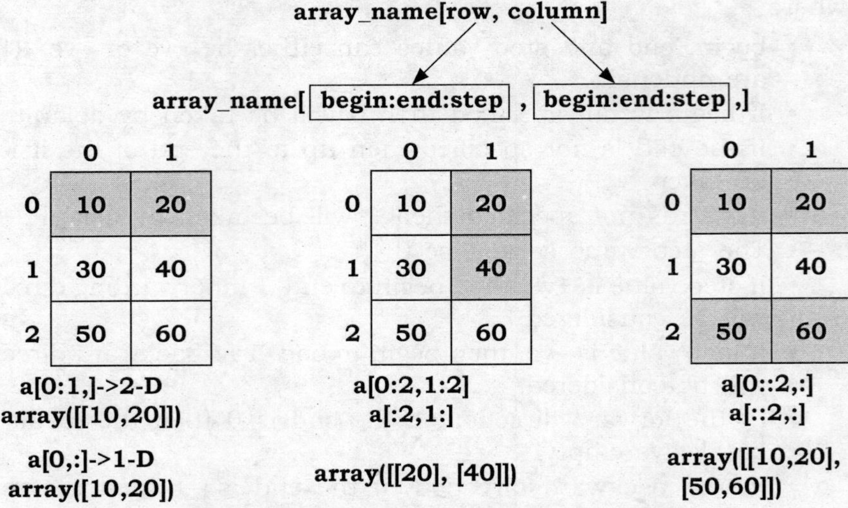

FIGURE 4.5   Slice operator on 2-D array

Program 10: A program to print values using Slice operator on 2-D array.

```
import numpy as np9
a = np9.array([[10,20],[30,40],[50,60]])
print(a)
```

OUTPUT

```
C:\h_numpy>py h9.py
[[10 20]
 [30 40]
 [50 60]]
```

Note:

- Row and column should be in slice syntax then only it returns 2-D array.
- If it is indexed, then it returns 1-D array.

Program 11: A program to print values using the Slice operator from the given array.

```
import numpy as np9
a = np9.array([[10,20],[30,40],[50,60]])
b = a[0,:] # row is in index form, it will return 1-D array
c = a[0:1,:] # it will return 2-D array
print("The dimension of b ",b.ndim and "the array b : ",b)
```

OUTPUT

```
C:\h_numpy>py h9.py
The dimension of b the array b: [10 20]
```

Program 12: A program to print values using the Slice operator from the given array.

```
import numpy as np9
a = np9.array([[10,20],[30,40],[50,60]])
print("a[0:2,1:2] value is : \n ",a[0:2,1:2])
print("a[:2,1:] value is : \n ",a[:2,1:])
```

OUTPUT

```
C:\h_numpy>py h9.py
a[0:2,1:2] value is :
[[20]
 [40]]
a[:2,1:] value is :
[[20]
 [40]]
```

Program 13: A program to print values using the Slice operator from the given array.

```
import numpy as np9
a = np9.array([[10,20],[30,40],[50,60]])
print("a[0::2,:] value is : \n ",a[0::2,:])
print("a[::2,:] value is : \n ",a[::2,:])
```

OUTPUT

```
C:\h_numpy>py h9.py
a[0::2,:] value is :
 [[10 20]
 [50 60]]
a[::2,:] value is :
 [[10 20]
 [50 60]]
```

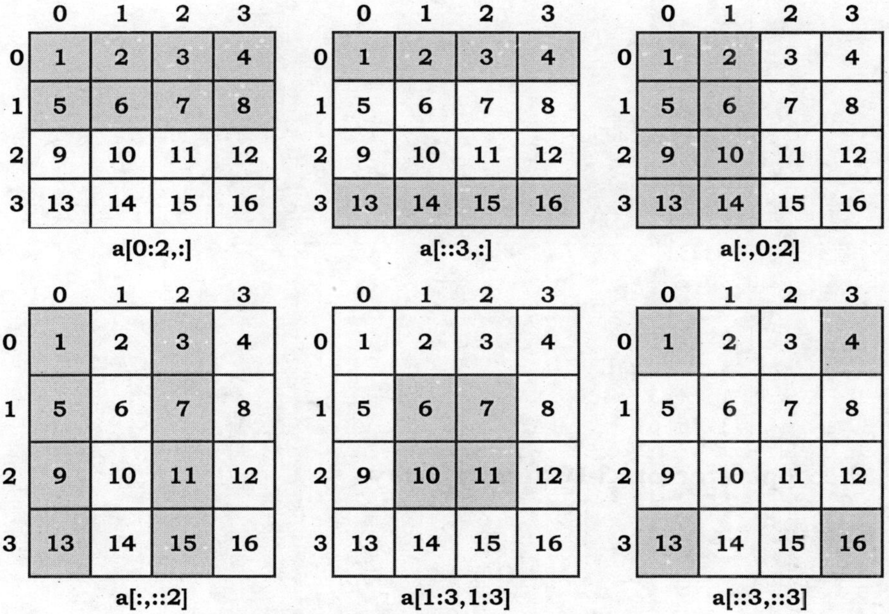

FIGURE 4.6 Slicing of 2-D array

Program 14: A program to print values of 4×4 array.

```
import numpy as np9
a = np9.arange(1,17).reshape(4,4)
print(a)
```

OUTPUT

```
C:\h_numpy>py h9.py
[[ 1  2  3  4]
 [ 5  6  7  8]
 [ 9 10 11 12]
 [13 14 15 16]]
```

Program 15: A program for applying slice operator on 4×4 array.

```
import numpy as np9
a = np9.arange(1,17).reshape(4,4)
print("a[0:2,:] value is : \n ",a[0:2,:])
print("a[0::3,:] value is : \n ",a[0::3,:])
print("a[:,:2] value is : \n ",a[:,:2])
print("a[:,::2] value is : \n ",a[:,::2])
print("a[1:3,1:3] value is : \n ",a[1:3,1:3])
print("a[::3,::3] value is : \n ",a[::3,::3])
```

OUTPUT

```
C:\h_numpy>py h9.py
a[0:2,:] value is :
 [[1 2 3 4]
 [5 6 7 8]]
a[0::3,:] value is :
 [[ 1  2  3  4]
 [13 14 15 16]]
a[:,:2] value is :
 [[ 1  2]
 [ 5  6]
 [ 9 10]
 [13 14]]
a[:,::2] value is :
 [[ 1  3]
 [ 5  7]
 [ 9 11]
 [13 15]]
a[1:3,1:3] value is :
 [[ 6  7]
 [10 11]]
a[::3,::3] value is :
 [[ 1  4]
 [13 16]]
```

Slice operator on 3-D NumPy array

Syntax

```
array_name[i,j,k]
```

where

i   indices of 2-D array (axis-0)

j   row index (axis-1)

k   column index (axis-2)

`array_name[begin:end:step,begin:end:step,begin:end:step]`

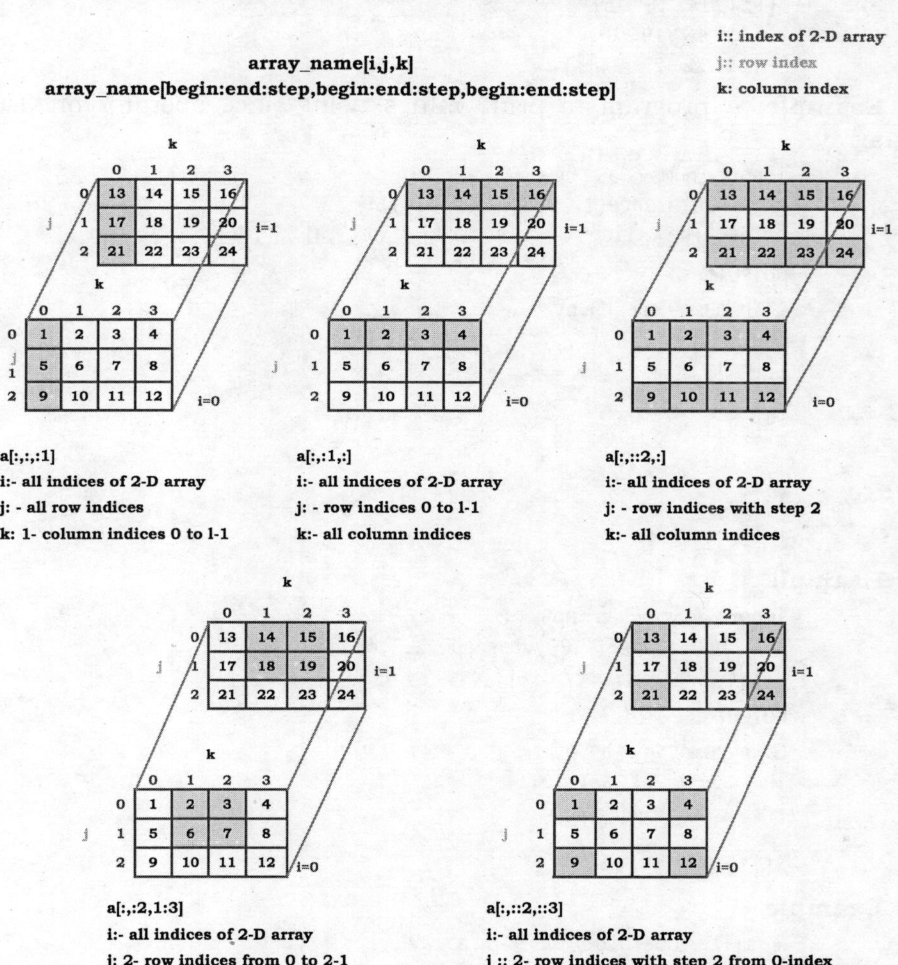

FIGURE 4.7   Slice operator on 3-D array

Program 16: A program to print values using Slice operator on 3-D array.

```
import numpy as np9
a = np9.arange(1,25).reshape(2,3,4)
print(a)
```

OUTPUT

```
C:\h_numpy>py h9.py
[[[ 1  2  3  4]
  [ 5  6  7  8]
  [ 9 10 11 12]]

 [[13 14 15 16]
  [17 18 19 20]
  [21 22 23 24]]]
```

Example: A program to print values using Slice operator on 3-D array.

```
import numpy as np9
a = np9.arange(1,25).reshape(2,3,4)
print("The slice of a[:,:,:1] :\n ",a[:,:,:1])
```

OUTPUT

```
C:\h_numpy>py h9.py
The slice of a[:,:,:1] :
[[[ 1]
  [ 5]
  [ 9]]

 [[13]
  [17]
  [21]]]
```

Example

```
import numpy as np9
a = np9.arange(1,25).reshape(2,3,4)
print("The slice of a[:,:1,:] \n ",a[:,:1,:])
```

OUTPUT

```
C:\h_numpy>py h9.py
The slice of a[:,:1,:]
[[[ 1  2  3  4]]

 [[13 14 15 16]]]
```

Example

```
# slice operator on 3-D array
import numpy as np9
a = np9.arange(1,25).reshape(2,3,4)
print("The slice of a[:,::2,:] \n ",a[:,::2,:])
```

OUTPUT

```
C:\h_numpy>py h9.py
The slice of a[:,::2,:]
[[[ 1  2  3  4]
  [ 9 10 11 12]]

 [[13 14 15 16]
  [21 22 23 24]]]
```

Example

```
# slice operator on 3-D array
import numpy as np9
a = np9.arange(1,25).reshape(2,3,4)
print(f"The slice of a[:,:2,1:3] \n ",a[:,:2,1:3])
```

OUTPUT

```
C:\h_numpy>py h9.py
The slice of a[:,:2,1:3]
  [[[ 2  3]
   [ 6  7]]

  [[14 15]
   [18 19]]]
```

Example

```
# slice operator on 3-D array
import numpy as np9
a = np9.arange(1,25).reshape(2,3,4)
print(f"The slice of a[:,::2,::3] \n ",a[:,::2,::3])
```

OUTPUT

```
C:\h_numpy>py h9.py
The slice of a[:,::2,::3]
  [[[ 1  4]
   [ 9 12]]

  [[13 16]
   [21 24]]]
```

Note:

- To use the Slice operator, it is necessary that the elements should be in order.
- We cannot select elements that are out of order, that is, we cannot select arbitrary elements.

4.3 ADVANCED INDEXING

It means a group of elements that are not in order (arbitrary elements). To access arbitrary elements (elements that are out of order), we should go for advanced indexing. By using an index, we can access only one element at a time. By using the Slice operator, we can access multiple elements at a time, but all elements should be in order.

Accessing multiple arbitrary elements in 1-D array

Syntax

```
array[x] ==> x can be either ndarray or list, which represents required indices
```

1st way (nd array)

Create an nd array with required indices in the given original array.
Pass that array as an argument to the original array.

2nd way (list)

Create a list with the required indices in the given original array.
Pass that list as an argument to the original array.

Example

```
import numpy as np9
a = np9.arange(10,101,10)
print(a)
```

OUTPUT

```
C:\h_numpy>py h9.py
[ 10 20 30 40 50 60 70 80 90 100]
```

1st way (using nd array)

```
# step1 : create an ndarray with the required indices
# assume that we have to extract 30,50,60,80 elements.
#Their indices are 2,4,5,8
# create an ndarray with those indices
import numpy as np9
ind = np9.array([2,4,5,8])
print(ind)
```

OUTPUT

```
C:\h_numpy>py h9.py
[2 4 5 8]
```

```
# step 2: pass the indices as argument to the original array
to get the required
#arbitrary elements
import numpy as np9
ind = np9.array([2,4,5,8])
a = np9.arange(10,101,10)
print(a[ind])
```

OUTPUT

```
C:\h_numpy>py h9.py
[30 50 60 90]
```

2nd way (using list)

```
# Step 1: create a list with the required indices of the
original array
import numpy as np9
a = np9.arange(10,101,10)
l = [2,4,5,8]
#Step 2: pass the list as an argument to the original array to
get the required #arbitrary elements
print(a[l])
```

OUTPUT

```
C:\h_numpy>py h9.py
[30 50 60 90]
```

Accessing multiple arbitrary elements in 2-D array

Syntax

```
a[[row_indices],[column_indices]]
```

a[[0,1,2,3],[0,1,2,3]   a[[0,1,2,3],[1,3,0,2]   a[[0,1,2,3,3,3,3],[0,0,0,0,1,2,3]

(0,0),(1,1),(2,2),(3,3)   (0,1),(1,3),(2,0),(3,2)   (0,0),(1,0),(2,0),(3,0),(3,1)
                                                                        (3,2),(3,3)

How to access the required elements from the 2-D array?

Step 1: Find the co-ordinates of the required elements (row, column).

Assume the required elements are 1, 5, 9, 13, 14, 15, 16

```
1 ::(0,0) 14 :: (3,1)
5 ::(1,0) 15::(3,2)
9 ::(2,0) 16::(3,3)
13 ::(3,0)
```

Step 2: Arrange them in order (0,0),(1,0),(2,0),(3,0),(3,1),(3,2),(3,3),

Step 3: Create row list with first co-ordinates [0,1,2,3,3,3,3].

Create column list with second co-ordinates [0,0,0,0,1,2,3].

Step 4: Pass these lists as argument to the original array

a[[0,1,2,3,3,3,3],[0,0,0,0,1,2,3]]

FIGURE 4.8 Accessing multiple arbitrary elements in 2-D array

Example

```
# It select elements from (0,0),(1,1),(2,2) and (3,3)
import numpy as np9
l= [[1,2,3,4],[5,6,7,8],[9,10,11,12],[13,14,15,16]]
a = np9.array(l)
print(a[[0,1,2,3],[0,1,2,3]])
```

OUTPUT

```
C:\h_numpy>py h9.py
[ 1  6  11  16]
```

Example

```
# It select elements from (0,0),(1,1),(2,2) and (3,3)
import numpy as np9
l= [[1,2,3,4],[5,6,7,8],[9,10,11,12],[13,14,15,16]]
a = np9.array(l)
print(a[[0,1,2,3],[1,3,0,2]])
```

OUTPUT

```
C:\h_numpy>py h9.py
[ 2  8  9  15]
```

Example

```
# IL-shape
import numpy as np9
l= [[1,2,3,4],[5,6,7,8],[9,10,11,12],[13,14,15,16]]
a = np9.array(l)
print(a[[0,1,2,3,3,3,3],[0,0,0,0,1,2,3]])
```

OUTPUT

```
C:\h_numpy>py h9.py
[ 1  5  9  13  14  15  16]
```

Note:

- In advanced indexing, the required elements will be selected based on indices and with those elements, a 1-D array will be created.
- The result of advanced indexing is always 1-D array even though we select 1-D or 2-D or 3-D array as input.
- But in slicing, the dimension of the output array is always the same as the dimension of input array.

Accessing multiple arbitrary elements in 3-D array

$$a[i][j][k]$$

where
   i   represents the index of 2-D array
   j   represents row index
   k   represents column index

Syntax

```
a[[indices of 2d array],[row indices],[column indices]]
```

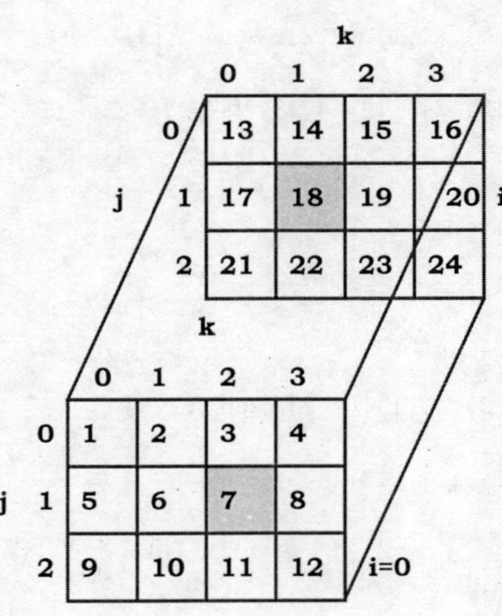

Assume that we have to select 7 and 18 elements

Step 1: Find the co-ordinates of arbitrary elements

7::	18::
i:0	i:1
j:0	j:1
k:2	k:1

Step 2: Create lists of i, j, k

i: [0,1]

j: [1:1]

k: [2,1]

Step 3: Pass these lists as arguments to the original array

a[[0,1],[1,1],[2,1]]

FIGURE 4.9 Accessing multiple arbitrary elements in 3-D array

Program 17: A program for accessing the arbitrary elements of 3-D arrays.

```
import numpy as np9
a = np9.arange(1,25).reshape(2,3,4)
print(a)
```

OUTPUT

```
C:\h_numpy>py h9.py
[[[ 1  2  3  4]
  [ 5  6  7  8]
  [ 9 10 11 12]]

 [[13 14 15 16]
  [17 18 19 20]
  [21 22 23 24]]]
```

Summary

- To access the arbitrary elements of 1-D array

    ```
    a[x] :: x can be either ndarray or list of indices
    ```

- To access the arbitrary elements of 2-D array

    ```
    a[[row_indices],[column_indices]]
    ```

- To access the arbitrary elements of 3-D array

    ```
    a[[indices of 2e-D array],[row_indices],[column_indices]]
    ```

4.4 CONDITION-BASED SELECTION

We can select elements of an array based on condition as well:

```
array_name[condition]
```

Syntax

```
array_name[boolean_array]
```

- In boolean_array, wherever True is present, the corresponding value will be selected.

Example

```
import numpy as np9
a = np9.array([10,20,30,40])
boolean_array=np9.array([True,False,False,True])
print(a[boolean_array])
```
OUTPUT
```
C:\h_numpy>py h9.py
[10 40]
```

Program 18: A program to select elements which are greater than 25 from the given array.

```
import numpy as np9
a = np9.array([10,20,30,40])  # boolean_array should be
[False,False,True,True]
```

```
# condition will give the boolean value
print(a>25)
```

OUTPUT

```
:\h_numpy>py h9.py
[False False True True]
```

Program 19: A program to print elements in single step to achieve the functionality (array_name[condition]).

```
import numpy as np9
a = np9.array([10,20,30,40])
print(a[a>25])
```

OUTPUT

```
C:\h_numpy>py h9.py
[30 40]
```

Program 20: A program to select negative numbers from the given array.

```
import numpy as np9
a = np9.array([10,-5,20,40,-3,-1,75])
print(a[a<0])
```

OUTPUT

```
C:\h_numpy>py h9.py
[-5 -3 -1]
```

Program 21: A program to select the even numbers from 2-D array.

```
import numpy as np9
a = np9.arange(1,26).reshape(5,5)
print(a[a%2==0])
```

OUTPUT

```
C:\h_numpy>py h9.py
[ 2  4  6  8  10  12  14  16  18  20  22  24]
```

Logical operations

- logical_and(x1, x2, /[, out, where, ...]) - Compute the truth value of x1 AND x2 element-wise.
- logical_or(x1, x2, /[, out, where, casting, ...]) - Compute the truth value of x1 OR x2 element-wise.
- logical_not(x, /[, out, where, casting, ...]) - Compute the truth value of NOT x element-wise.
- logical_xor(x1, x2, /[, out, where, ...]) - Compute the truth value of x1 XOR x2, element-wise.

Program 22: A program to select elements by using multiple conditions by using logical operations which are even and divisible by 5.

```
import numpy as np9
a = np9.arange(1,26).reshape(5,5)
b = np9.logical_and(a%2==0,a%5==0)
```

```
print("Original array a ",a)
print("Logical array b ",b)
print("Elements which are even and divisible by 5 ",a[b])
```

OUTPUT

```
C:\h_numpy>py h9.py
Original array a [[ 1  2  3  4  5]
 [ 6  7  8  9 10]
 [11 12 13 14 15]
 [16 17 18 19 20]
 [21 22 23 24 25]]
Logical array b [[False False False False False]
 [False False False False  True]
 [False False False False False]
 [False False False False  True]
 [False False False False False]]
Elements which are even and divisible by 5 [10 20]
```

Selecting elements based on multiple conditions

We can use & for AND condition and | for OR condition.

Syntax

```
array_name[(condition_1) & (condition_2)]
array_name[(condition_1) | (condition_2)]
```

Program 23: A program to select the elements which are divisible by 2 and divisible by 3.

```
import numpy as np9
a = np9.arange(1,26).reshape(5,5)
print(a[(a%2 == 0) & (a%3 ==0)])
```

OUTPUT

```
C:\h_numpy>py h9.py
[ 6 12 18 24]
```

Program 24: A program to select the elements with boolean combination.

```
import numpy as np9
a = np9.arange(1,26).reshape(5,5)
bool_array = np9.logical_and(a%2==0,a%5==0)
print(a[bool_array])
```

OUTPUT

```
C:\h_numpy>py h9.py
[10 20]
```

Program 25: A program to select either the even or odd elements which are divisible by 3.

```
import numpy as np9
a = np9.arange(1,26).reshape(5,5)
print(a[np9.logical_or(a%2==0,a%3==0)])
```

```
OUTPUT
C:\h_numpy>py h9.py
[ 2  3  4  6  8  9  10  12  14  15  16  18  20  21  22  24]
```

Program 26: A program to select the elments which are divisible by 2 and 3.

```
import numpy as np9
a = np9.arange(1,26).reshape(5,5)
print(a[(a%2 == 0) & (a%3 ==0)])
```

```
OUTPUT
C:\h_numpy>py h9.py
[ 6  12  18  24]
```

Program 27: A program to select the elments which are divisible by either 2 or 3.

```
import numpy as np9
a = np9.arange(1,26).reshape(5,5)
print(a[(a%2 == 0) | (a%3 ==0)])
```

```
OUTPUT
C:\h_numpy>py h9.py
[ 2  3  4  6  8  9  10  12  14  15  16  18  20  21  22  24]
```

4.5 PYTHON SLICING VS NUMPY ARRAY SLICING VS ADVANCED INDEXING VS CONDITION-BASED SELECTION

Case-1: Python Slicing

- In the case of list, Slice operator will create a **separate copy**.
- If we perform any change in one copy, those changes won't be reflected in other copy.

Example

```
# Python Slicing ==> new copy is created
l1 = [10,20,30,40]
l2=l1[::]
print(l1 is l2)
```

```
OUTPUT
C:\h_numpy>py h9.py
False
```

Example

```
# if we made changes in l1 it does not reflect on l2 and vice-
versa
l1 = [10,20,30,40]
l2=l1[::]
l1[0] = 888
print("l1::",l1)
print("l2::",l2)
```

OUTPUT
```
C:\h_numpy>py h9.py
l1:: [888, 20, 30, 40]
l2:: [10, 20, 30, 40]
```

Case-2: NumPy Array Slicing

In NumPy array slicing, a separate copy won't be created and we get a view of the original copy. The view is a logical entity, whereas Table is physical entity (RDBMS).

Program 28: A program to print elements from NumPy array using slicing.

```
import numpy as np9
a = np9.arange(10,101,10)
print(a)
```

OUTPUT
```
C:\h_numpy>py h9.py
[ 10 20 30 40 50 60 70 80 90 100]
```

Program 29: A program to print the elements using slicing on NumPy array.

```
import numpy as np9
a = np9.arange(10,101,10)
b = a[0:4] # view is created
print("b ",b)
print("a ",a)
```

OUTPUT
```
C:\h_numpy>py h9.py
b [10 20 30 40]
a [ 10 20 30 40 50 60 70 80 90 100]
```

Case-3: Advanced Indexing and Condition-based Selection

It selects required elements based on provided index or condition. With those elements, a new 1-D array object will be created. The output is always a new 1-D array.

Program 30: A program to print elements of an array using advanced indexing.

```
import numpy as np9
a = np9.arange(10,101,10)
b = a[[0,2,5]]
print(" a ",a)
print(" b ",b)
```

OUTPUT
```
C:\h_numpy>py h9.py
a [ 10 20 30 40 50 60 70 80 90 100]
b [10 30 60]
```

Slicing vs Advanced Indexing

Slicing

- The elements should be ordered.
- We can't select arbitrary elements.
- The condition-based selection is not possible.
- We get an original view but not a separate copy.
- It is best in terms of performance.

Advanced indexing

- The elements need not to be ordered.
- We can select arbitrary elements.
- Condition-based selection is possible.
- We get a separate copy.
- Its performance is not upto the mark as compared to slicing.

EXERCISES

1. Write a program to know the position of an element in the 4-D array.

2. Write a program to print values using Slice operator on entire array in reverse order.

3. Write a program for accessing 7 and 18 from the array.

4. Write a program to check whether the condition-based selection is applicable for 2-D array also.

5. Write a program to check if changes are made in b, will it be reflected on a and vice-versa?

5 ARITHEMETIC OPERATORS

5.1 INTRODUCTION

The following are the various arithmetic operators:

    +   Addition
    -    Subtraction
    *   Multiplication
    /   Division
    //   Floor division
    %   Modulo operation/Remainder operation
    **   Exponential operation/Power operation

Note:

- The result of division operator (/) is always float.
- The floor division operator (//) returns either integer and float values.
- If both arguments are of int type then the floor division operator returns int value only.
- If at least one argument is float type then it returns float type only.

```
print("10/2 value :: ",10/2)
print("10.0/2value :: ",10.0/2)
print("10//2 value :: ",10//2)
print("10.0//2 value :: ",10.0//2)
```

OUTPUT

```
C:\h_numpy>py h9.py
10/2 value :: 5.0
10.0/2value :: 5.0
10//2 value :: 5
10.0//2 value :: 5.0
```

5.2 ARITHMETIC OPERATORS FOR NUMPY ARRAYS WITH SCALAR

Scalar means constant numeric value. All arithmetic operators are applicable for NumPy arrays with a scalar. All arithmetic operations will be performed at the element level.

Program 1: A program for arithmetic operations using the 1-D array.

```
import numpy as np9
a = np9.array([10,20,30,40])
print("a+2 value is :: ",a+2)
print("a-2 value is :: ",a-2)
print("a*2 value is :: ",a*2)
print("a**2 value is :: ",a**2)
print("a/2 value is :: ",a/2)
print("a//2 value is :: ",a//2)
```

OUTPUT

```
C:\h_numpy>py h9.py
a+2 value is :: [12 22 32 42]
a-2 value is :: [ 8 18 28 38]
a*2 value is :: [20 40 60 80]
a**2 value is :: [ 100 400 900 1600]
a/2 value is :: [ 5. 10. 15. 20.]
a//2 value is :: [ 5 10 15 20]
```

Program 2: A program for arithmetic operations using the 2-D array.

```
import numpy as np9
a = np9.array([[10,20,30],[40,50,60]])
print("a value is ::\n ",a)
print("a+2 value is :: \n ",a+2)
print("a-2 value is :: \n ",a-2)
print("a*2 value is :: \n ",a*2)
print("a**2 value is ::\n ",a**2)
print("a/2 value is :: \n ",a/2)
print("a//2 value is ::\n ",a//2)
```

OUTPUT

```
C:\h_numpy>py h9.py
a value is ::
[[10 20 30]
[40 50 60]]
a+2 value is ::
[[12 22 32]
 [42 52 62]]
a-2 value is ::
    [[ 8 18 28]
    [38 48 58]]
```

```
a*2 value is ::
   [[ 20 40 60]
    [ 80 100 120]]
a**2 value is ::
   [[ 100 400 900]
    [1600 2500 3600]]
a/2 value is ::
   [[ 5.  10.  15.]
    [20. 25. 30.]]
a//2 value is ::
   [[ 5 10 15]
    [20 25 30]]
```

5.3 ZeroDivisionError

In Python, when a value is divided by zero (zero/zero), it results in ZeroDivisionError. But in NumPy, there is no ZeroDivisionError.

```
10/0 ==> Infinity(inf)
0/0 ==> undefined(nan--->not a number)
```

The numbers that are divisible by 0 results infinity.

```
print("The value of 10/0 :: ",10/0)
print("The value of 0/0 :: ",0/0)
```

OUTPUT

```
C:\h_numpy>py h9.py
Traceback (most recent call last):
File "C:\h_numpy\h9.py", line 2, in <module>
print("The value of 10/0 :: ",10/0)
ZeroDivisionError: division by zero
```

Program 3: A program to check the result when a value is divided by zero.

```
import numpy as np9
a = np9.arange(6)
print("The value of a/0 :: ",a/0)
```

OUTPUT

```
C:\h_numpy\h9.py:4: RuntimeWarning: divide by zero encountered
in true_divide
print("The value of a/0 :: ",a/0)
C:\h_numpy\h9.py:4: RuntimeWarning: invalid value encountered in
true_divide
print("The value of a/0 :: ",a/0)
The value of a/0 :: [nan inf inf inf inf inf]
```

5.4 ARITHMETIC OPERATORS FOR ARRAYS WITH ARRAYS

To perform arithmetic operations between NumPy arrays, both arrays should have same dimensions, shapes and sizes, otherwise we will get an error.

Program 4: A program to perform arithmetic operations on the 1-D array.

```
import numpy as np9
a = np9.array([1,2,3,4])
b = np9.array([10,20,30,40])
print("Dimension of a : ",a.ndim," size of a :",a.shape," and shape of a :
    ",a.shape)
print("Dimension of b : ",b.ndim, "size of b :",b.shape," and shape of b :
    ",b.shape)
print("a array :: ",a," and b array :: ",b)
print("a+b value is :: ",a+b)
print("a-b value is :: ",a-b)
print("a*b value is :: ",a*b)
print("a**b value is :: ",a**b)
```

OUTPUT

```
C:\h_numpy>py h9.py
Dimension of a : 1 size of a : (4,) and shape of a : (4,)
Dimension of b : 1 size of b : (4,) and shape of b : (4,)
a array :: [1 2 3 4] and b array :: [10 20 30 40]
a+b value is :: [11  22  33  44]
a-b value is :: [-9 -18 -27 -36]
a*b value is :: [10  40  90 160]
a**b value is :: [1 1048576 -1010140999 0]
```

Program 5: A program to perform arithmetic operations on the 2-D array.

```
import numpy as np9
a = np9.array([[1,2],[3,4]])
b = np9.array([[5,6],[7,8]])
print("Dimension of a : ",a.ndim," size of a :",a.shape, "and shape of a :
    ",a.shape)
print("Dimension of b : ",b.ndim," size of b :",b.shape, "and shape of b :
    ",b.shape)
print("a array :: \n ",a)
print("b array :: \n ",b)
print("a+b value is :: \n ",a+b)
print("a-b value is :: \n ",a-b)
print("a*b value is :: \n ",a*b)
print("a**b value is :: \n ",a**b)
print("a/b value is :: \n ",a/b)
print("a//b value is :: \n ",a//b)
```

OUTPUT

```
C:\h_numpy>py h9.py
Dimension of a : 2 size of a : (2, 2) and shape of a : (2, 2)
Dimension of b : 2 size of b : (2, 2) and shape of b : (2, 2)
a array ::
 [[1 2]
  [3 4]]
```

```
b array ::
 [[5 6]
  [7 8]]
a+b value is ::
 [[ 6  8]
  [10 12]]
a-b value is ::
 [[-4 -4]
  [-4 -4]]
a*b value is ::
 [[ 5 12]
  [21 32]]
a**b value is ::
 [[ 1 64]
  [ 2187 65536]]
a/b value is ::
 [[0.2 0.33333333]
  [0.42857143 0.5  ]]
a//b value is ::
 [[0 0]
  [0 0]]
```

5.5   FUNCTIONS OF ARITHMETIC OPERATORS IN NUMPY

The equivalent functions of arithmetic operators in NumPy are as follows:

```
a+b        np9.add(a,b)
a-b        np9.subtract(a,b)
a*b        np9.multiply(a,b)
a/b        np9.divide(a,b)
a//b       np9.floor_divide(a,b)
a%b        np9.mod(a,b)
a**b       np9.power(a,b)
```

Note: To use arithmetic functions, both arrays should be of the same dimensions, sizes, and shapes.

5.6   UNIVERSAL FUNCTIONS (ufunc)

The function which operates element by element on the whole array is called universal function (ufunc). All the functions listed above are ufunc.

```
np9.dot()        Matrix Multiplication/Dot product
np9.multiply()   Element multiplication
```

EXERCISES

1. Write a program to perform arithmetic operations using arithmetic functions.

6 BROADCASTING

6.1 INTRODUCTION

Generally, the arithmetic operations are performed between two arrays that are having the same dimensions, shapes, and sizes. Even though dimensions, shapes and sizes are different, still some arithmetic operations are allowed by broadcasting. Broadcasting will be performed automatically by NumPy itself and it is not required to perform explicitly. Broadcasting won't be possible in all cases.

Note:

- If both arrays have the same dimensions, shapes, and sizes then broadcasting is not required.
- In case of different dimensions, shapes or sizes, broadcasting is required.

6.2 RULES OF BROADCASTING

NumPy follows some rules to perform broadcasting. If the rules are satisfied then only broadcasting will be performed internally for performing arithmetic operations.

Rule-1: Make sure both arrays should have same dimensions

If the two arrays are of different dimensions, NumPy will make equal dimensions by adding 1s on the left side of the lesser dimension array, until both arrays have the same dimensions.

Example

Before

    a array has shape :: (4, 3) :::: 2-D array
    b array has shape :: (3,) :::: 1-D array

After

- Both arrays a and b are of different dimensions.
- From Rule-1, NumPy will add 1s to the lesser dimension array (here array b).

Now,

- The array a is (4, 3) and the array b becomes ::: (1, 3) (By using Rule-1)
- Both arrays are of same dimensions (array a:: 2-D and array-b :: 2-D)

  a array has shape :: (4, 3) :::: 2-D array
  b array has shape :: (1, 3) :::: 2-D array

Rule-2

If the size of two arrays does not match in any dimension, then the arrays with a size equal to 1 in that dimension will be increased to the size of other dimensions to match.

Example

- From Rule-1, we got a ==> (4, 3) and b ==> (1, 3)
- Here first co-ordinate of a => 4 and b => 1. Thus the sizes are different.

Example

Let us find out whether the broadcasting between (3, 2, 2) and (3,) is possible or not.

Before applying Rule-1

    a :: (3, 2, 2) – 3-D array
    b :: (3,) – 1-D array

Here both are in different dimensions. By applying Rule-1, the NumPy changes the array b as (1, 1, 3).

After applying Rule-1

    a :: (3, 2, 2) – 3-D array
    b :: (1, 1, 3) – 3-D array

Now both arrays are in the same dimensions.

Before applying Rule-2

    a :: (3, 2, 2) – 3-D array
    b :: (1, 1, 3) – 3-D array

By applying Rule-2, NumPy changes the array b to (3, 2, 3). It will be replaced with the corresponding sizes of array a.

After applying Rule-2

    a :: (3, 2, 2) – 3-D array
    b :: (3, 2, 3) – 3-D array

Now, arrays a and b have the same dimensions, but different shapes. So NumPy is unable to perform broadcasting.

According to Rule-2, the array with size 1 (here array b with size 1) will be increased to 4 (corresponding size of array a). The second co-ordinate of a => 3 and b => 3. Thus, the sizes are matched. After applying Rule-2, the dimensions of a and b are changed as follows:

```
a array has shape ==> (4, 3) ====> 2-D array
b array has shape ==> (4, 3) ====> 2-D array
```

Now, both the arrays have same dimensions, shapes and sizes. So we can perform any arithmetic operations.

Note:

- In any dimension, the sizes are not matched and neither equal to 1, then we will get an error. NumPy is not able to perform broadcasting between such arrays.
- The data will be reused from the same input array.
- If the rows are required then reuse existing rows.
- If columns are required then reuse existing columns.
- The result is always a higher dimension of input arrays.

PRACTICE PROGRAMS

Program 1: A program to perform arithmetic operations between 1-D arrays of different sizes using broadcasting.

```
import numpy as np9
a = np9.array([10,20,30,40])
b = np9.array([1,2,3])
# a : (4,) b: (3,).
# Both are having same dimensions. But sizes are different so arithmetic operation is not performed
# Broadcasting by NumPy will be failed while performing arithmetic operation in  this case
print(a+b)
```

OUTPUT

```
C:\h_numpy>py h9.py
Traceback (most recent call last):
  File "C:\h_numpy\h9.py", line 8, in <module>
    print(a+b)
ValueError: operands could not be broadcast together with shapes (4,) (3,)
```

Program 2: A program to apply broadcasting between 1-D arrays.

```
import numpy as np9
a = np9.array([10,20,30])
```

```
b = np9.array([40])
print("Shape of_a : ",a.shape)
print("Shape of b : ",b.shape)
print("a+b : ",a+b)
```

OUTPUT

```
C:\h_numpy>py h9.py
Shape of a : (3,)
Shape of b : (1,)
a+b : [50 60 70]
```

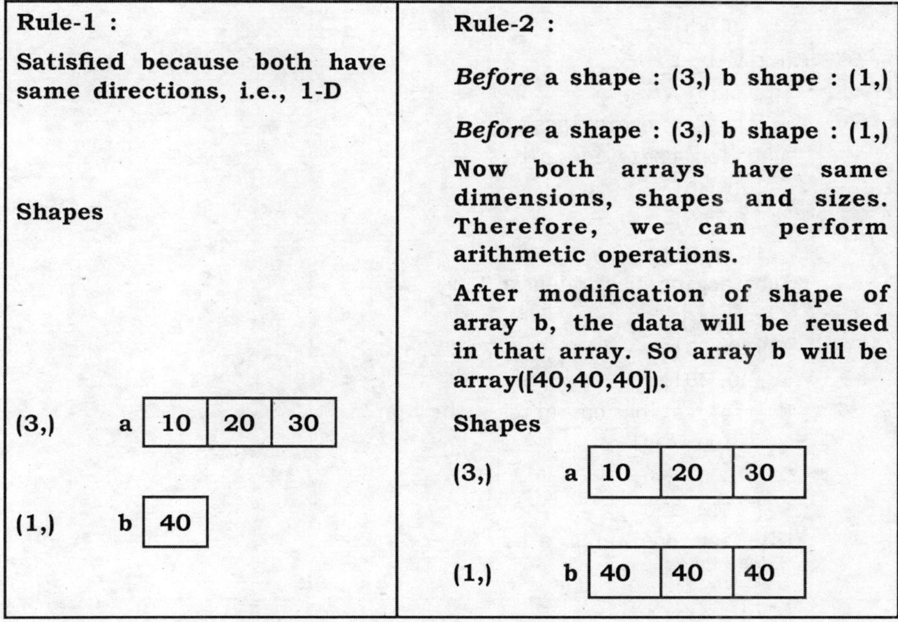

FIGURE 6.1   Broadcasting of 1-D Array

Program 3: A program to apply broadcasting between 2-D and 1-D arrays.

```
import numpy as np9
a = np9.array([[10,20],[30,40],[50,60]])
b = np9.array([10,20])
print("Shape of the array a : ",a.shape)
print("Shape of the array b : ",b.shape)
print("array a :\n ",a)
print("array b :\n ",b)
print("Arithmetic operations :")
print("Addition operation a+b :\n ",a+b)
print("Subtraction operation a-b :\n ",a-b)
print("Multiplication operation a*b:\n ",a*b)
```

```python
print("Division operation a/b:\n ",a/b)
print("Floor division operation a//b:\n .",a//b)
print("Modulo operation a%b:\n ",a%b)
```

OUTPUT

```
C:\h_numpy>py h9.py
Shape of the array a : (3, 2)
Shape of the array b : (2,)
array a :
 [[10 20]
  [30 40]
  [50 60]]
array b :
 [10 20]
Arithmetic operations :
Addition operation a+b :
 [[20 40]
  [40 60]
  [60 80]]
Subtraction operation a-b :
 [[ 0 0]
  [20 20]
  [40 40]]
Multiplication operation a*b:
 [[ 100 400]
  [ 300 800]
  [ 500 1200]]
Division operation a/b:
 [[1. 1.]
  [3. 2.]
  [5. 3.]]
Floor division operation a//b:
 [[1 1]
  [3 2]
  [5 3]]
Modulo operation a%b:
 [[0 0]
  [0 0]
  [0 0]]
```

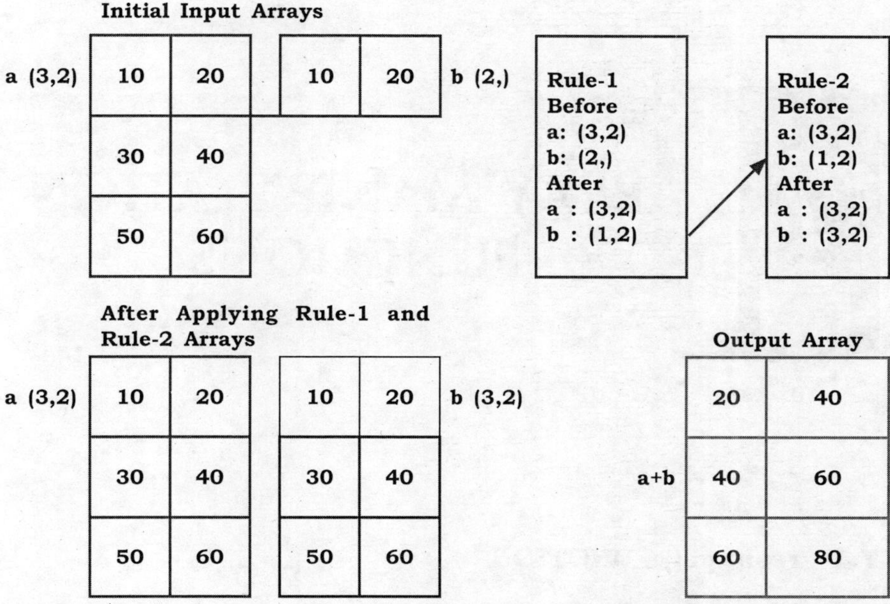

FIGURE 6.2   Broadcasting using Rule 1 and Rule 2 Arrays

EXERCISES

1. Write a program to perform arithematic operations using broadcasting between 1-D and 2-D arrays.

7 ARRAY MANIPULATION FUNCTIONS

7.1 reshape() FUNCTION

We can use reshape() function to change array shape without changing data.

Syntax

```
reshape(a, newshape, order='C')
```

It gives a new shape to an array without changing its data.

```
newshape : int or tuple of ints
order : {'C', 'F', 'A'}
    Default value for the order: 'C'
    'C' ==> C language style which means row major order.
    'F' ==> Fortran language style which means column major order.
```

The data should not be changed. Input size and output size should be matched, otherwise we will get the ValueError. As there is no change in the data, so new objects won't be created. We will get just View. If we perform any change in the original array, that change will be reflected in reshaped array and vice versa. We can specify unknown dimension size as -1, but only once.

```
reshape(a, newshape, order='C')
```

- C style ==> Row major order (default): it will consider 1st row, 2nd row and so on
- Fortran Style (F) ==> Column major order: it will consider 1st column, 2nd column and so on

We can use either reshape() function of the NumPy array or reshape() method of nd array.

```
np9.reshape(a,new_shape) ==> Functional style
ndarray_obj.reshape(new_shape) ==> Object oriented style
# one shape to any other shape
# (10,) --> (5,2),(2,5),(10,1),(1,10) ==> any no. of views
# (24,) --> (3,8), (6,4),(2,3,4), (2,2,2,4) all are valid
```

The nd array class also contains reshape() method and hence we can call this method on any nd array object.

- numpy ==> module
- nd array ==> class present in NumPy module
- reshape() ==> method present in ndarray class
- reshape() ==> function present in NumPy module.

numpy.reshape() ==> numpy library function

b = np9.reshape (a,shape,order)

ndarray.reshape()---->ndarray object method

b = a.reshape (shape,order)

Program 1: A program to check that the sizes of the arrays should be matched otherwise we will get error using reshape().

```
import numpy as np9
a = np9.arange(1,11)
print("array : ",a)
b = np9.reshape(a,(5,3))
```

OUTPUT

```
C:\h_numpy>py h9.py
array : [ 1 2 3 4 5 6 7 8 9 10]
Traceback (most recent call last):
 File "C:\h_numpy\h9.py", line 5, in <module>
   b = np9.reshape(a,(5,3))
 File "<__array_function__ internals>", line 5, in reshape
 File "C:\Users\SureshS\AppData\Local\Programs\Python\Python39\lib\
 site-packages\numpy\core\fromnumeric.py", line 298, in reshape
 return _wrapfunc(a, 'reshape', newshape, order=order)
 File "C:\Users\Suresh S\AppData\Local\Programs\Python\Python39\
 lib\ site-packages\numpy\core\fromnumeric.py", line 57, in _
 wrapfunc
 return bound(*args, **kwds)
ValueError: cannot reshape array of size 10 into shape (5,3)
```

Program 2: A program to convert 1-D array to 2-D array.

```
import numpy as np9
a = np9.arange(1,11)
print("array : ",a)
b = np9.reshape(a,(5,2))
print("Converting 1-D(a) array to 2-D(b) array : \n ",b)
```

OUTPUT

```
C:\h_numpy>py h9.py
array : [ 1 2 3 4 5 6 7 8 9 10]
```

```
Converting 1-D(a) array to 2-D(b) array :
 [[ 1  2]
  [ 3  4]
  [ 5  6]
  [ 7  8]
  [ 9 10]]
```

By using ndarray_obj.reshape() ==> Object oriented style

```
import numpy as np9
a = np9.arange(1,25)
b = a.reshape((2,3,4))
print("*"*80)
print("Original array : \n ",a)
print("Reshaped array :\n ",b)
print("*"*80)
```

OUTPUT

```
C:\h_numpy>py h9.py
********************************************************************************
Original array :
[  1  2  3  4  5  6  7  8  9 10 11 12 13 14 15 16 17 18 19 20 21
22 23 24]
Reshaped array :
 [[[ 1  2  3  4]
   [ 5  6  7  8]
   [ 9 10 11 12]]
  [[13 14 15 16]
   [17 18 19 20]
   [21 22 23 24]]]
```

order='C'

- C style :: row major order
- It will consider 1st row, 2nd row and so on

Program 3: A program to print an array using C style order (Row major order).

```
import numpy as np9
a = np9.arange(12)
b = a.reshape((3,4))
print("*"*80)
print("Original array : ",a)
print("Reshaped array :\n ",b)
print("*"*80)
```

OUTPUT

```
C:\h_numpy>py h9.py
********************************************************************************
Original array : [ 0  1  2  3  4  5  6  7  8  9 10 11]
Reshaped array :
 [[ 0  1  2  3]
  [ 4  5  6  7]
  [ 8  9 10 11]]
```

order='F'

- Fortran style :: column major order
- It will consider 1st column, 2nd column and so on

Program 4: A program to print an array using Fortran style order (Column major order).

```
import numpy as np9
a = np9.arange(12)
b = a.reshape((3,4),order='F')
print("*"*80)
print("Original array : ",a)
print("Reshaped array :\n ",b)
print("*"*80)
```

OUTPUT

```
C:\h_numpy>py h9.py
********************************************************************************
Original array : [ 0  1  2  3  4  5  6  7  8  9 10 11]
Reshaped array :
 [[ 0  3  6  9]
 [ 1  4  7 10]
 [ 2  5  8 11]]
```

Unknown dimension with –1

We can use an unknown dimension with –1, only once. NumPy will decide and replace –1 with the required dimension. We can use any -ve number instead of –1. One shape dimension can be –1. In this case, the value is inferred from the length of the array and the remaining dimensions are inferred automatically by NumPy itself.

```
b = a.reshape((5,-1))   #valid
b = a.reshape((-1,5))   #valid
b = a.reshape((-1,-1))  #invalid
```

Example

```
import numpy as np9
a = np9.arange(1,11)
b = a.reshape((5,-1))
print("Shape of array b :: ",b.shape)
```

OUTPUT

```
C:\h_numpy>py h9.py
Shape of array b :: (5, 2)
```

Example

```
import numpy as np9
a = np9.arange(1,11)
b = a.reshape((-1,5))
print("Shape of array b :: ",b.shape)
```

OUTPUT

```
C:\h_numpy>py h9.py
Shape of array b :: (2, 5)
```

Example

```
import numpy as np9
a = np9.arange(1,11)
b = a.reshape((-1,-1))
print("Shape of array b :: ",b.shape)
```

OUTPUT

```
C:\h_numpy>py h9.py
Traceback (most recent call last):
  File "C:\h_numpy\h9.py", line 3, in <module>
    b = a.reshape((-1,-1))
ValueError: can only specify one unknown dimension
```

Conclusion of reshape() function

- It reshapes the array without changing data.
- The sizes must be matched.
- We can use either NumPy library function(np9.reshape()) or ndarray class method(a.reshape()).
- It won't create a new array object, just we will get a view.
- We can use −1 in the case of an unknown dimension, but only once.

7.2 resize()

It returns a new array with a specified size.

Syntax

```
numpy.resize(a, new_shape)
```

where

- Output array can be of any dimension, shape, and size.
- Input size and output size need not to be matched.
- The data may be changed.
- We will get the repeated copies.
- We can get the new data by np9.resize() and a.resize().
- The unknown dimension −1 is not applicable in this resize().

By using np9.resize() ==> Functional Style

Syntax

```
numpy.resize(a, new_shape)
```

- If new_size requires more elements, repeat the elements of the input array.
- The new object will be created.

Example

```
import numpy as np9
a = np9.arange(1,6)
```

```
b = np9.resize(a,(2,4))
print("Original array : ",a)
print("Reshaped array :\n ",b)
```

OUTPUT

```
C:\h_numpy>py h9.py
Original array : [1 2 3 4 5]
Reshaped array :
 [[1 2 3 4]
  [5 1 2 3]]
```

Example

```
import numpy as np9
a= np9.arange(1,6)
c = np9.resize(a,(2,2))
print(" array c : \n ",c)
```

OUTPUT

```
C:\h_numpy>py h9.py
 array c :
 [[1 2]
  [3 4]]
```

Program 5: A program to print original array a that is not modified when we call np9.resize() function.

```
import numpy as np9
a= np9.arange(1,6)
c = np9.resize(a,(2,2))
print(a)
```

OUTPUT

```
C:\h_numpy>py h9.py
[1 2 3 4 5]
```

By using ndarray_obj.resize() ==> Object oriented style

Syntax

```
ndarray_object.resize(new_shape, refcheck=True)
```

- If new_size requires more elements then extra elements are filled with zeros.
- If we are using the nd array class resize() method, the existing array will be modified.

Example

```
import numpy as np9
m = np9.arange(1,6)
print("Original array :\n ",m)
m.resize((4,2))
print("After calling resize() method in ndarray :\n ",m)
```

OUTPUT

```
C:\h_numpy>py h9.py
```

```
Original array :
[1 2 3 4 5]
After calling resize() method in ndarray :
[[1 2]
 [3 4]
 [5 0]
 [0 0]]
```

Difference between numpy.resize() and ndarray.resize()

Table 7.1 numpy.resize() vs ndarray.resize()

numpy.resize()	ndarray.resize()
It is the library function present in numpy module.	It is the method present in ndarray class.
It will create a new array.	It won't return new array and existing array will be modified (inline modification).
If the new_shape requires more elements then repeated copies of original array will be used.	If the new_shape requires more elements then extra elements are filled with zeros.

Difference between reshape() and resize()

Table 7.2 rehape() vs resize()

reshape()	resize()
It changes the shape of array, but not the size.	It changes the size of the array. The shape and data may be changed automatically.
It won't create new array object and we will get view of existing array.	It will create new array object with required new object.
If we perform any change in the reshaped copy, those changes will be reflected in the original copy automatically and vice versa.	If we perform any change in the resized array, those changes won't be reflected in the original copy.
There will be no change in the original data in reshape().	There may be a chance of data change in resize() [either expansion or shrinking].
The size must be matched.	The size need not to be matched.
In unknown dimension, we can use −1.	Type −1 is not applicable.

7.3 flatten() METHOD

We can use flatten() method to flatten (convert) any n-dimensional array to 1-dimensional array.

Syntax

```
ndarray.flatten(order='C')
```

It creates a new 1-dimensional array with elements of given n-dimensional array, i.e., Return a copy of the array collapsed in one dimension.

```
C-style====>row major order
F-stype===>column major order
```

It is the method present in nd array class but not in the NumPy library function.

```
a.flatten()--->valid
np9.flatten()-->invalid
```

It creates a new array and returns it (i.e., copy, but not the view). If any change is made in the original array, it won't be reflected in flatten copy and vice versa. The output of flatten method is always the 1-D array.

Program 6: A program to convert 2-D array to 1-D array using C style.

```
import numpy as np9
a = np9.arange(6).reshape(3,2)
b = a.flatten()
print("Original array :\n ",a)
print("Flatten array :\n ",b)
```

OUTPUT

```
C:\h_numpy>py h9.py
Original array :
 [[0 1]
 [2 3]
 [4 5]]
Flatten array :
 [0 1 2 3 4 5]
```

Program 7: A program to convert a 2-D array to 1-D array in Fortran style F.

```
import numpy as np9
a = np9.arange(6).reshape(3,2)
b = a.flatten('F')
print("Original array :\n ",a)
print("Flatten array :\n ",b)
```

OUTPUT

```
C:\h_numpy>py h9.py
Original array :
 [[0 1]
 [2 3]
 [4 5]]
Flatten array :
 [0 2 4 1 3 5]
```

Program 8: A program to convert a 3-D to 1-D array using C style.

```
import numpy as np9
a = np9.arange(1,19).reshape(3,3,2)
b = a.flatten()
print("Original array :\n ",a)
print("Flatten array :\n ",b)
```

OUTPUT

```
C:\h_numpy>py h9.py
Original array :
 [[[ 1  2]
   [ 3  4]
   [ 5  6]]

  [[ 7  8]
   [ 9 10]
   [11 12]]

  [[13 14]
   [15 16]
   [17 18]]]
Flatten array:
 [ 1  2  3  4  5  6  7  8  9 10 11 12 13 14 15 16 17 18]
```

Program 9: A program to convert a 3-D to 1-D in Fortran style F.

```
a = np9.arange(1,19).reshape(3,3,2)
b = a.flatten('F')
print("Original array :\n ",a)
print("Flatten array :\n ",b)
```

OUTPUT

```
C:\h_numpy>py h9.py
Original array :
 [[[ 1  2]
   [ 3  4]
   [ 5  6]]

  [[ 7  8]
   [ 9 10]
   [11 12]]

  [[13 14]
   [15 16]
   [17 18]]]
Flatten array :
 [ 1  7 13  3  9 15  5 11 17  2  8 14  4 10 16  6 12 18]
```

7.4 flat VARIABLE

It is the 1-D iterator over the array. This is a 'numpy.flatiter' instance.

```
ndarray.flat    → Return a flat iterator over an array.
ndarray.flatten → Returns a flattened copy of an array.
```

Example

```
import numpy as np9
a = np9.arange(1,7).reshape(3,2)
print(a.flat)
```

OUTPUT

```
C:\h_numpy>py h9.py
<numpy.flatiter object at 0x00000247191272B0>
```

Example

```
import numpy as np9
a = np9.arange(1,7).reshape(3,2)
print(a.flat[4])
```

OUTPUT

```
C:\h_numpy>py h9.py
5
```

Iterate over flat for x in a.flat: print(x)

Example: # iterate over flat

```
import numpy as np9
a = np9.arange(1,7).reshape(3,2)
for x in a.flat: print(x)
```

OUTPUT

```
C:\h_numpy>py h9.py
1
2
3
4
5
6
```

7.5 ravel()

It is a library function in NumPy module and a method in the nd array class. It is exactly the same as flatten function except that it returns a view but not a copy. This function is used to convert any n-D array to a 1-D array. It is a method which is present in nd array class and the NumPy library function.

```
np9.ndarray.ravel() → valid
np9.ravel() → valid
```

a.ravel(order='C')

```
C-style → row major order
F-stype → column major order
```

It returns a view but not a copy. If we make a change in the original array then the changes will be reflected in the ravel copy also.

Note: The output of ravel() method is always the 1-D array.

Program 10: A program to arrange array using ravel() method in ndarray class.

```
import numpy as np9
a = np9.arange(24).reshape(2,3,4)
b = a.ravel()
print("Original array :\n ",a)
print("Ravel array :\n ",b)
OUTPUT
C:\h_numpy>py h9.py
Original array :
 [[[ 0  1  2  3]
  [ 4  5  6  7]
  [ 8  9 10 11]]
 [[12 13 14 15]
  [16 17 18 19]
  [20 21 22 23]]]
Ravel array :
 [ 0  1  2  3  4  5  6  7  8  9 10 11 12 13 14 15 16 17 18 19 20
 21 22 23]
```

Program 11: A program to print ravel array using assign value.

```
import numpy as np9
a = np9.arange(24).reshape(2,3,4)
b = a.ravel()
b[0] = 8888
print("Ravel array :\n ",b)
print("Original array :\n ",a)

OUTPUT

C:\h_numpy>py h9.py
Ravel array :
 [8888  1  2  3  4  5  6  7  8  9 10 11 12 13
 14 15 16 17 18 19 20 21 22 23]
Original array :
 [[[8888  1  2  3]
  [ 4  5  6  7]
  [ 8  9 10 11]]

 [[ 12 13 14 15]
  [ 16 17 18 19]
  [ 20 21 22 23]]]
```

Difference between flatten() and ravel()

Table   7.3 flatten() vs ravel()

flatten()	ravel()
It can be used to flatten n-D array to 1-D array. A separate 1-D array object will be created.	It can be used to flatten n-D array to 1-D array, but it won't create a new 1-D array object and we will get only view.

If we perform any changes in the flatten() copy, then those changes won't be reflected in the original copy.	If we perform any changes in the ravel() copy, then those changes will be reflected in the original copy.
flatten() method operates slower than ravel() as it creates a new array object.	ravel() method operates faster than flatten() as it does not create a new array object, and it just returns a view.
flatten() is not NumPy library level function and it is a method present in nd array class. `ndarrayObj.flatten()` → valid `numpy.flatten(a)` → invalid	ravel() is both NumPy library level function and nd array class method. `ndarrayObj.ravel()` → valid `numpy.ravel(a)` → valid

7.6 transpose()

Generally, we get transpose of a given matrix by interchanging rows and columns. Similarly, to interchange dimensions of nd array, we use transpose() function.

Syntax

```
numpy.transpose(a,axes=None)
```

- Reverse or permute the axes of an array; returns the modified array (view but not copy).
- For an array with two axes, transpose(a) gives the matrix transpose.
- If we are not providing axes argument value then the dimensions are simply reversed.
- In transpose() operation, only dimensions will be interchanged, but not content. Hence it is not required to create a new array. Thus it returns a view of the existing array.

Example: Shapes in the dimension

```
(2,3)--->(3,2)
(2,3,4)-->(4,3,2),(2,4,3),(3,2,4),(3,4,2)
```

$$A = \begin{bmatrix} 1 & 2 \\ 3 & 4 \end{bmatrix} \quad A^T = \begin{bmatrix} 1 & 3 \\ 2 & 4 \end{bmatrix}$$

FIGURE 7.1   Transpose of A

Example

```
import numpy as np9
a = np9.array([[1,2],[3,4]])
atrans = np9.transpose(a)
print("Original Array : \n ",a)
print("Transposed Array : \n ",atrans)
```

```
OUTPUT
C:\h_numpy>py h9.py
Original Array :
 [[1 2]
  [3 4]]
Transposed Array :
 [[1 3]
  [2 4]]
```

When we call the transpose() function without specifying the axes parameter then input dimension will be simply reversed.

```
1-D array (4,) ==> Transposed Array ::(4,) No change
2-D array (2,3) ==> Trasposed Array :: (3,2)
3-D array (2,3,4) ==> Transposed Array :: (4,3,2)
4-D array (2,2,3,4)==> Transposed Array :: (4,3,2,2)
```

If we are not specifying the axes parameter, dimensions will be reversed

Original Array Shape	Transpose Array Shape
1-D :: (4,)	**(4,) No change**
2-D :: (2,3)	**(3,2)**
3-D :: (2,3,4)	**(4,3,2)**
4-D :: (2,2,3,4,)	**(4,3,2,2)**

FIGURE 7.2 Transpose dimensions of array

Program 12: A program to print a transpose of 1-D array.

```
import numpy as np9
a = np9.array([10,20,30,40])
atrans = np9.transpose(a)
print("Original Array : \n ",a)
print("Transposed Array : \n ",atrans)
print("Original Array shape : ",a.shape)
print("Transposed Array shape: ",atrans.shape)
```

```
OUTPUT
C:\h_numpy>py h9.py
Original Array :
 [10 20 30 40]
Transposed Array :
 [10 20 30 40]
Original Array shape : (4,)
Transposed Array shape: (4,)
```

3-D array shape : (2,3,4):

```
2--->2 2-Dimensional arrays
3 --->In every 2-D array, 3 rows
4 --->In every 2-D array, 4 columns
24--->Total number of elements
```

If we transpose the 3-D array then the shape will be (4,3,2):

```
4 ---> 4 2-D arrays
3 --->In every 2-D array, 3 rows
2 --->In every 2-D array, 2 columns
24--->Total number of elements
```

Program 13: A program to print a transpose of 4-D array.

```python
import numpy as np9
a = np9.arange(1,49).reshape(2,2,3,4)
atrans = np9.transpose(a)
print("Original Array : \n ",a)
print("Transposed Array : \n ",atrans)
print("Original Array shape : ",a.shape)
print("Transposed Array shape: ",atrans.shape)
```

OUTPUT

```
C:\h_numpy>py h9.py
Original Array :
[[[[ 1  2  3  4]
   [ 5  6  7  8]
   [ 9 10 11 12]]

  [[13 14 15 16]
   [17 18 19 20]
   [21 22 23 24]]]


 [[[25 26 27 28]
   [29 30 31 32]
   [33 34 35 36]]

  [[37 38 39 40]
   [41 42 43 44]
   [45 46 47 48]]]]
Transposed Array :
[[[[ 1 25]
   [13 37]]

  [[ 5 29]
   [17 41]]

  [[ 9 33]
   [21 45]]]


 [[[ 2 26]
   [14 38]]

  [[ 6 30]
   [18 42]]

  [[10 34]
   [22 46]]]
```

```
[[[  3  27]
  [15  39]]

 [[  7  31]
  [19  43]]

 [[11  35]
  [23  47]]]


[[[  4  28]
  [16  40]]

 [[  8  32]
  [20  44]]

 [[12  36]
  [24  48]]]]
Original Array shape : (2, 2, 3, 4)
Transposed Array shape: (4, 3, 2, 2)
```

Specifying the 'axes' parameter while calling the transpose() function

- Axes parameter describes in which order we have to take axes.
- It is very helpful in 3-D and 4-D arrays.
- We can specify the customized dimensions.

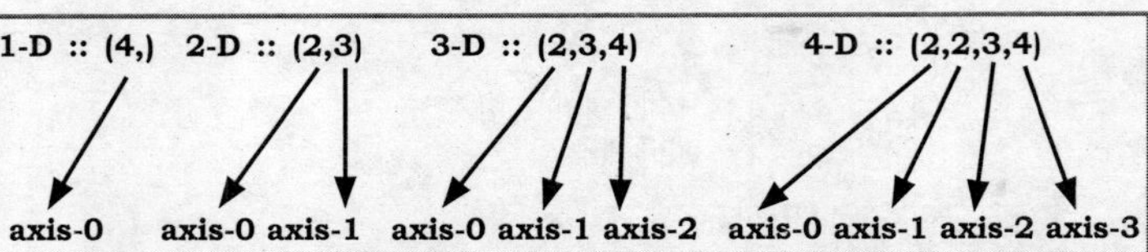

Original Array Shape	Transpose Array Shape
1-D :: (4,)	(4,)
2-D :: (2,3)	(3,2) --> axes=(1,0) or axes=(0,1)
3-D :: (2,3,4)	(2,4,3),(3,2,4),(3,4,2),(4,3,2),(4,2,3) (2,4,3)-->axes=(0,2,1)(3,4,2) -->axes=(1,2,0) (3,2,4)-->axes=(0,2,1) (4,3,2) -->axes=(2,1,0) (4,3,2) -->axes=(2,0,1)

FIGURE 7.3　Transpose array shape

Program 14: A program to print transpose of 1-D array with axes parameter.

```
import numpy as np9
a = np9.array([10,20,30,40])
atrans = np9.transpose(a,axes=0)
print("Original Array : \n ",a)
```

```
print("Transposed Array : \n ",atrans)
print("Original Array shape : ",a.shape)
print("Transposed Array shape: ",atrans.shape)
```

OUTPUT

```
C:\h_numpy>py h9.py
Original Array :
 [10 20 30 40]
Transposed Array :
 [10 20 30 40]
Original Array shape : (4,)
Transposed Array shape: (4,)
```

Program 15: A program to print transpose of 2-D array with axes parameter.

```
import numpy as np9
a = np9.arange(1,7).reshape(2,3)
atrans = np9.transpose(a,axes=(0,1))
print("Original Array : \n ",a)
print("Transposed Array : \n ",atrans)
print("Original Array shape : ",a.shape)
print("Transposed Array shape: ",atrans.shape)
```

OUTPUT

```
C:\h_numpy>py h9.py
Original Array :
 [[1 2 3]
  [4 5 6]]
Transposed Array :
 [[1 2 3]
  [4 5 6]]
Original Array shape : (2, 3)
Transposed Array shape: (2, 3)
```

Program 16: A program to print transpose 3-D array [transpose()] with axes parameter.

```
# (2,3,4) ==> Original Array shape
# 2--->2-D arrays (axis-0 : 0)
# 3--->3 rows in every 2-D array (axis-1 : 1)
# 4--->4 columns in every 2-D array (axis-2 : 2)
# (2,4,3) ==> Customized Transposed array shape
        import numpy as np9
        a = np9.arange(1,25).reshape(2,3,4)
        atrans = np9.transpose(a,axes=(0,2,1))
        print("Original Array : \n ",a)
        print("Transposed Array : \n ",atrans)
        print("Original Array shape : ",a.shape)
        print("Transposed Array shape: ",atrans.shape)
```

OUTPUT

```
C:\h_numpy>py h9.py
```

```
Original Array :
[[[ 1   2   3   4]
  [ 5   6   7   8]
  [ 9  10  11  12]]

 [[13  14  15  16]
  [17  18  19  20]
  [21  22  23  24]]]
Transposed Array :
[[[ 1  5   9]
  [ 2  6  10]
  [ 3  7  11]
  [ 4  8  12]]

 [[13  17  21]
  [14  18  22]
  [15  19  23]
  [16  20  24]]]
Original Array shape: (2, 3, 4)
Transposed Array shape: (2, 4, 3)
```

Note: If we repeat the same axis multiple times then we will get error.

transpose() ::

It is the nd array class method. The behaviour is exactly the same as the NumPy library function transpose().

Program 17: A program to print 3-D array using tanspose().

```python
import numpy as np9
a = np9.arange(1,25).reshape(2,3,4)
atrans = a.transpose((0,2,1))
print("Original Array : \n ",a)
print("Transposed Array : \n ",atrans)
print("Original Array shape : ",a.shape)
print("Transposed Array shape: ",atrans.shape)
```

OUTPUT

```
C:\h_numpy>py h9.py
Original Array :
 [[[ 1 2 3 4]
  [ 5 6 7 8]
  [ 9 10 11 12]]

 [[13 14 15 16]
  [17 18 19 20]
  [21 22 23 24]]]
Transposed Array :
 [[[ 1 5  9]
  [ 2 6 10]
  [ 3 7 11]
  [ 4 8 12]]
```

```
       [[13  17  21]
        [14  18  22]
        [15  19  23]
        [16  20  24]]]
Original Array shape : (2, 3, 4)
Transposed Array shape: (2, 4, 3)
```

T :: variable in ndarray class

We can use shortcut representation of the transpose with T variable.

```
•  ndarray_obj.T ==> it will simply reverse the dimensions
•  Customized dimensions are possible with T variable
```

Program 18: A program to print transpose array using T variable.

```
import numpy as np9
a = np9.arange(1,25).reshape(2,3,4)
atrans = a.T
print("Original Array : \n ",a)
print("Transposed Array : \n ",atrans)
print("Original Array shape : ",a.shape)
print("Transposed Array shape: ",atrans.shape)
```

OUTPUT

```
C:\h_numpy>py h9.py
Original Array :
 [[[ 1   2   3   4]
   [ 5   6   7   8]
   [ 9  10  11  12]]

  [[13  14  15  16]
   [17  18  19  20]
   [21  22  23  24]]]
Transposed Array :
 [[[ 1  13]
   [ 5  17]
   [ 9  21]]

  [[ 2  14]
   [ 6  18]
   [10  22]]

  [[ 3  15]
   [ 7  19]
   [11  23]]

  [[ 4  16]
   [ 8  20]
   [12  24]]]
Original Array shape : (2, 3, 4)
Transposed Array shape: (4, 3, 2)
```

reshape vs transpose

In reshape, we can change the size of dimension, but the total size remains the same.

Example

```
Input:  (3,4)
Output: (4,3),(2,6),(6,2),(1,12),(12,1),(2,2,3),(3,2,2)
```

But in transpose, the dimensions are interchanged. However, there is no change in the size of any dimension.

Example

```
Input: (3,4) ==> Output: (4,3)
```

Example

```
Input: (2,3,4) ==> Output: (4,3,2),(2,4,3),(3,2,4),(3,4,2) but we
cannot take(2,12),(3,8). Transpose is special case of reshape
```

Various possible syntaxes for transpose():

```
numpy.transpose(a)
numpy.transpose(a,axes=(2,0,1))
ndarrayobject.transpose()
ndarrayobject.transpose(*axes)
ndarrayobject.T
```

Note: 1, 3, 5 lines are equal w.r.t., functionality.

Program 19: A program to print a transpose of the 1-D array using swapaxes.

```
import numpy as np9
a = np9.arange(1,7).reshape(3,2)
aswap = np9.swapaxes(a,0,1)
print("Original Array : \n ",a)
print("Transposed Array : \n ",aswap)
print("Original Array shape : ",a.shape)
print("Transposed Array shape: ",aswap.shape)
```

OUTPUT

```
C:\h_numpy>py h9.py
Original Array :
 [[1 2]
 [3 4]
 [5 6]]
Transposed Array :
 [[1 3 5]
 [2 4 6]]
Original Array shape : (3, 2)
Transposed Array shape: (2, 3)
```

Program 20: A program to print a transpose of the 2-D array using swapaxes.

```
import numpy as np9
a = np9.arange(1,7).reshape(3,2)
b = np9.swapaxes(a,0,1)
c = np9.swapaxes(a,1,0)
```

```
print("Swapped Array b : \n ",b)
print("Swapped Array c : \n ",c)
print("Swapped Array b shape : ",b.shape)
print("Swapped Array c shape : ",c.shape)
```

OUTPUT

```
C:\h_numpy>py h9.py
Swapped Array b :
 [[1 3 5]
  [2 4 6]]
Swapped Array c :
 [[1 3 5]
  [2 4 6]]
Swapped Array b shape : (2, 3)
Swapped Array c shape : (2, 3)
```

7.7 swapaxes()

It is a NumPy library function and nd array class method. By using transpose, we can interchange any number of dimensions. But if we want to interchange only two dimensions then we should go for the swapaxes() function. It is the special case of transpose() function.

Program 21: A program to print 3-D array using swapaxes function.

```
import numpy as np9
a = np9.arange(1,25).reshape(2,3,4)
aswap = a.swapaxes(0,1) # 0 and 1 axes values will be swapped
 : (3,2,4)
print("Original Array : \n ",a)
print("Original Array shape : ",a.shape)
print("Swapped Array shape: ",aswap.shape)
```

OUTPUT

```
C:\h_numpy>py h9.py
Original Array :
 [[[ 1  2  3  4]
   [ 5  6  7  8]
   [ 9 10 11 12]]

  [[13 14 15 16]
   [17 18 19 20]
   [21 22 23 24]]]
Original Array shape : (2, 3, 4)
Swapped Array shape: (3, 2, 4)
```

transpose() vs swapaxes()

- By using tarnspose(), we can interchange any number of dimensions.
- With swapaxes(), we can interchange only two dimensions.

EXERCISES

1. Write a program to check whether any change in the original array will be reflected to the reshaped array.

2. Write a program to change the value of original array and the changes won't be reflected in the flatten array.

3. Write a program to print an array after reshape using ravel() function in NumPy module.

4. Write a program to print transpose of 2-D array (In 2-D array, because of transpose, rows will become columns and columns will become rows.)

5. Write a program to print 3-D array after swaping using swapaxes.

8   JOINING AND SPLITTING OF ARRAYS

8.1   INTRODUCTION

It is similar to join queries in Oracle. We can join/concatenate multiple nd arrays into a single array by using the following functions:

1. concatenate()
2. stack()
3. vstack()
4. hstack()
5. dstack()

1. concatenate()

This function joins a sequence of arrays along an existing axis.

Syntax

```
concatenate((a1, a2, ...)), axis=0, out=None, dtype=None,
casting="same_kind")
```

where

(a1, a2,..)	input arrays along an existing axis
axis	based on which axis we have to perform concatenation axis=0(default) :: vertical concatenation will happens axis=1 :: Horizontal concatenation will happens axis=None :: First the arrays will be flatten (converted to 1-D array) and then concatenation will be performed on the resultant arrays.
out	destination array, where we have to store concatenation result.

Program 1: A program to print 1-D array using concatenate () function.

```
import numpy as np9
a = np9.arange(4)
b = np9.arange(5)
c = np9.arange(3)
print(np9.concatenate((a,b,c)))
```

OUTPUT

```
C:\h_numpy>py h9.py
[0 1 2 3 0 1 2 3 4 0 1 2]
```

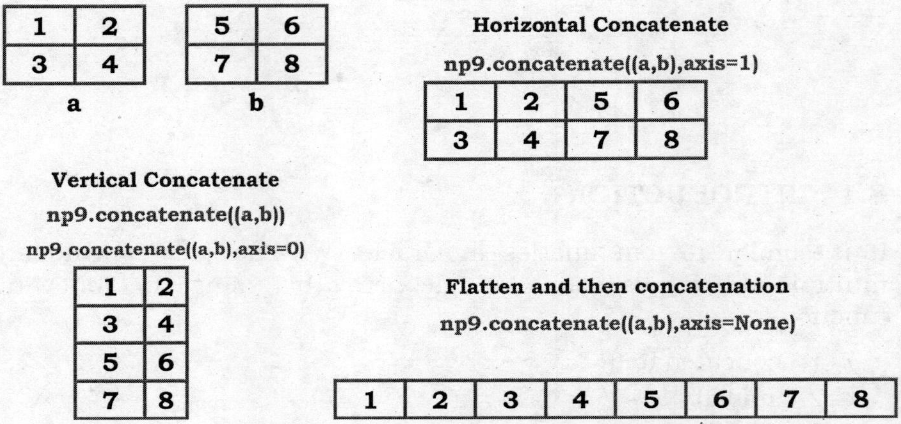

FIGURE 8.1   Concatenation of 1-D array

Program 2: A program to print 2-D array using concatenate () function.

```
import numpy as np9
a = np9.array([[1,2],[3,4]])
b = np9.array([[5,6],[7,8]])
# concatenation by providing axis parameter
# Vertical Concatenation
vcon = np9.concatenate((a,b))
vcon1 = np9.concatenate((a,b),axis=0)
# Horizontal Concateation
hcon = np9.concatenate((a,b),axis=1)
# flatten and then concatenation
flatt = np9.concatenate((a,b),axis=None)
print(f"array a ",a)
print(f"array b ",b)
print(f"Without specifying axis parameter ",vcon)
print(f"Specifying axis=0 a ",vcon1)
```

```
print(f"Specifying  axis=1  ",hcon)
print(f"Specifying  axis=None  ",flatt)
```

OUTPUT

```
C:\h_numpy>py h9.py
array a [[1 2]
         [3 4]]
array b [[5 6]
         [7 8]]
Without specifying axis parameter [[1 2]
                                   [3 4]
                                   [5 6]
                                   [7 8]]
Specifying axis=0 a [[1 2]
                     [3 4]
                     [5 6]
                     [7 8]]
Specifying axis=1 [[1 2 5 6]
                   [3 4 7 8]]
Specifying axis=None [1 2 3 4 5 6 7 8]
```

Rules

- We can join any number of arrays, but all arrays should be of the same dimensions.
- The sizes of all axes, except concatenation axis should be same.
- The shapes of resultant array and out array must be same.

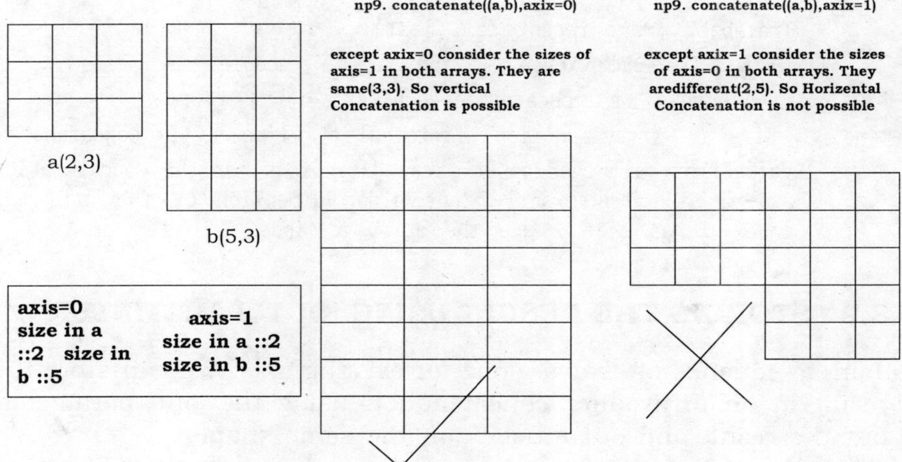

FIGURE 8.2   Array concatenation using axes

Program 3: A program to print 3-D array using concatenation ()
function (Rule-2 demonstration).

```
import numpy as np9
a = np9.arange(6).reshape(2,3)
b = np9.arange(15).reshape(5,3)
print("array a ",a)
print("array b ",b)
# axis=0 ==> Vertical concatenation
vcon = np9.concatenate((a,b),axis=0)
print("Vertical Concatenation array ", vcon)
# axis=1 ==> Horizontal Concatenation
hcon = np9.concatenate((a,b),axis=1)
print("Horizontal Concatenation array ",hcon)
```

OUTPUT

```
C:\h_numpy>py h9.py
array a [[0 1 2]
         [3 4 5]]
array b [[ 0  1  2]
         [ 3  4  5]
         [ 6  7  8]
         [ 9 10 11]
         [12 13 14]]
Vertical Concatenation array [[ 0  1  2]
                              [ 3  4  5]
                              [ 0  1  2]
                              [ 3  4  5]
                              [ 6  7  8]
                              [ 9 10 11]
                              [12 13 14]]
Traceback (most recent call last):
  File "C:\h_numpy\h9.py", line 11, in <module>
    hcon = np9.concatenate((a,b),axis=1)
  File "<__array_function__ internals>", line 5, in concatenate
ValueError: all the inp9ut array dimensions for the concatenation
axis must match exactly, but along dimension 0, the array at
index 0 has size 2 and the array at index 1 has size 5
```

8.2   STORING THE RESULT USING 'OUT' PARAMETER

Join a sequence of arrays along an existing axis. We can store the
result in an array after concatenation using the 'out' parameter,
but the result and out must be in the same shape.

Syntax

```
concatenate((a1, a2, ...), axis=0, out=None, dtype=None, casting="same_
kind")
```

Program 4: A program to print an array using out parameter.

```
import numpy as np9
a = np9.arange(4)
b = np9.arange(5)
c = np9.empty(9) # default dtype for empty is float
print(np9.concatenate((a,b),out=c))
```

OUTPUT

```
C:\h_numpy>py h9.py
[0. 1. 2. 3. 0. 1. 2. 3. 4.]
```

Using 'dtype' parameter

We can specify the required data type using dtype parameter.

Program 5: A program to print dtype parameter values.

```
import numpy as np9
a = np9.arange(4)
b = np9.arange(5)
print(np9.concatenate((a,b),dtype=str))
```

OUTPUT

```
C:\h_numpy>py h9.py
['0' '1' '2' '3' '0' '1' '2' '3' '4']
```

Note:

- We can use either out or dtype.
- We cannot use both out and dtype simultaneously because out has its own data type.

Program 6: A program to print for both out and dtype parameters.

```
import numpy as np9
a = np9.arange(4)
b = np9.arange(5)
c = np9.empty(9,dtype=str)
print(np9.concatenate((a,b),out=c,dtype=str))
```

OUTPUT

```
C:\h_numpy>py h9.py
Traceback (most recent call last):
  File "C:\h_numpy\h9.py", line 6, in <module>
print(np9.concatenate((a,b),out=c,dtype=str))
  File "<__array_function__ internals>", line 5, in concatenate
TypeError: concatenate() only takes 'out' or 'dtype' as an
argument, but both were provided.
```

Concatenation of 1-D arrays

We can concatenate any number of 1-D arrays at a time. For 1-D arrays, there will be only one axis, i.e., axis-0.

Program 7: A program to print the concatenation of three 1-D arrays.

```
import numpy as np9
a = np9.arange(4)
b = np9.arange(5)
c = np9.arange(3)
print(np9.concatenate((a,b,c),dtype=int))
```

OUTPUT

```
C:\h_numpy>py h9.py
[0 1 2 3 0 1 2 3 4 0 1 2]
```

Concatenation of 2-D arrays

We can concatenate any number of 2-D arrays at a time. For 2-D arrays, there will be two axes, i.e., axis-0 and axis-1.

where

   axis-0 ==> represents number of rows

   axis-1 ==> represents number of columns

We can perform concatenation on either axis-0 or axis-1. The size of all dimensions (axes) must be matched except concatenation axis.

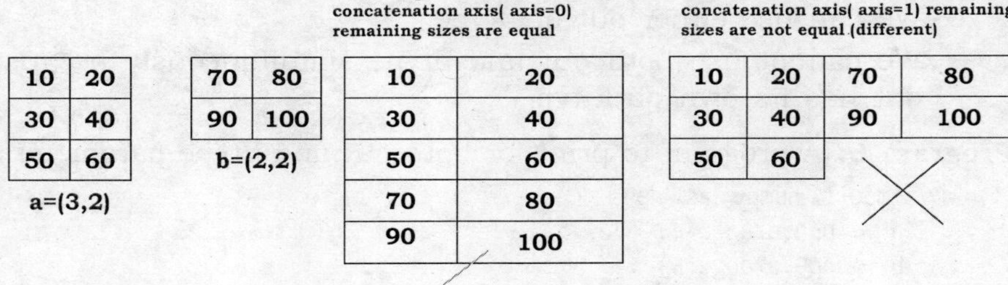

FIGURE 8.3   Concatenation of 2-D arrays

Concatenation of 3-D arrays

We can concatenate any number of 3-D arrays at a time. For 3-D arrays, there will be three axes, i.e., axis-0, axis-1 and axis-2.

   axis-0 ==> represents number of 2-D arrays

   axis-1 ==> represents number of rows in every 2-D array

   axis-2 ==> represents number of columns in every 2-D array

We can perform concatenation on axis-0, axis-1, axis-2 (existing axis). The size of all dimensions (axes) must be matched except concatenation axis.

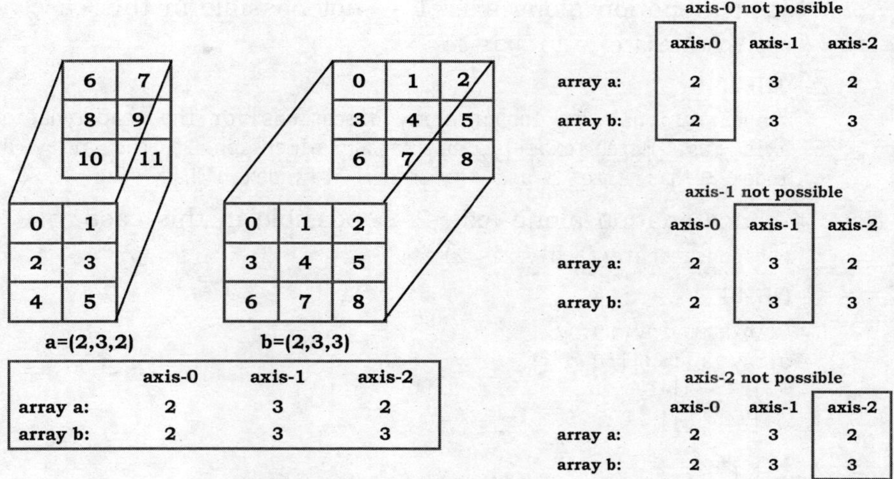

FIGURE 8.4   Concatenation of 3-D arrays

Program 8: A program to print an array after concatenation of 3-D arrays.

```
import numpy as np9
a = np9.arange(12).reshape(2,3,2)
b = np9.arange(18).reshape(2,3,3)
print("array a : ",a)
print("array b : ",b)
# concatenation along axis=0 ==> not possible in this case
print(np9.concatenate((a,b),axis=0))
```

OUTPUT

```
C:\h_numpy>py  h9.py
array a : [[[ 0  1]
            [ 2  3]
            [ 4  5]]

           [[ 6  7]
            [ 8  9]
            [10 11]]]
array b : [[[ 0  1  2]
            [ 3  4  5]
            [ 6  7  8]]

           [[ 9 10 11]
            [12 13 14]
            [15 16 17]]]
Traceback (most recent call last):
  File "C:\h_numpy\h9.py", line 8, in <module>
    print(np9.concatenate((a,b),axis=0))
  File "<__array_function__ internals>", line 5, in concatenate
ValueError: all the inp9ut array dimensions for the concatenation
axis must match exactly, but along dimension 2, the array at
index 0 has size 2 and the array at index 1 has size 3
```

Note: # concatenation along axis=1 → not possible in this case

```
np9.concatenate((a,b),axis=1)
```

OUTPUT

```
ValueError: all the inp9ut array dimensions for the concatenation
axis must match exactly, but along dimension 2, the array at
index 0 has size 2 and the array at index 1 has size 3
```

Note: # concatenation along axis=2 → possible in this case

```
np9.concatenate((a,b),axis=2)
```

OUTPUT

```
C:\h_numpy>py h9.py
array a : [[[ 0  1]
  [ 2   3]
  [ 4   5]]

 [[ 6    7]
  [ 8    9]
  [10  11]]]
array b : [[[ 0  1  2]
  [ 3  4  5]
  [ 6  7  8]]

 [[ 9  10  11]
  [12  13  14]
  [15  16  17]]]

[[[ 0  1  0  1  2]
  [ 2  3  3  4  5]
  [ 4  5  6  7  8]]

 [[ 6   7   9  10  11]
  [ 8   9  12  13  14]
  [10  11  15  16  17]]]
```

Program 9: A program to print an array after concatenation of 3-D array in all axes.

```
import numpy as np9
a = np9.arange(18).reshape(2,3,3)
b = np9.arange(18,36).reshape(2,3,3)
print("array a : ",a)
print("array b : ",b)
print("array a shape: ",a.shape)
print("array b shape: ",b.shape)
```

OUTPUT

```
C:\h_numpy>py h9.py
array a : [[[ 0  1  2]
  [ 3  4  5]
  [ 6  7  8]]
```

```
       [[ 9 10 11]
        [12 13 14]
        [15 16 17]]]
array b : [[[18 19 20]
        [21 22 23]
        [24 25 26]]
       [[27 28 29]
        [30 31 32]
        [33 34 35]]]
array a shape: (2, 3, 3)
array b shape: (2, 3, 3)
```


FIGURE 8.5   Concatenation of 4-D arrays

Program 10: A program to print an array after concatenation along axis-0.

```
import numpy as np9
a = np9.arange(18).reshape(2,3,3)
b = np9.arange(18,36).reshape(2,3,3)
axis0_result = np9.concatenate((a,b),axis=0)
print("Concatenation along axis-0 : ",axis0_result)
print("Shape of the resultant array : ",axis0_result.shape)
OUTPUT
C:\h_numpy>py h9.py
Concatenation along axis-0 : [[[ 0  1  2]
  [ 3  4  5]
  [ 6  7  8]]

 [[ 9 10 11]
  [12 13 14]
  [15 16 17]]

 [[18 19 20]
  [21 22 23]
  [24 25 26]]

 [[27 28 29]
  [30 31 32]
  [33 34 35]]]
Shape of the resultant array : (4, 3, 3)
```

Program 11: A program to print an array after concatenation along axis-2.

```
import numpy as np9
a = np9.arange(18).reshape(2,3,3)
b = np9.arange(18,36).reshape(2,3,3)
axis2_result = np9.concatenate((a,b),axis=2)
print("Concatenation along axis-2 : \n ",axis2_result)
print("Shape of the resultant array : ",axis2_result.shape)
OUTPUT
C:\h_numpy>py h9.py
Concatenation along axis-2 :
[[[ 0  1  2 18 19 20]
  [ 3  4  5 21 22 23]
  [ 6  7  8 24 25 26]]

 [[ 9 10 11 27 28 29]
  [12 13 14 30 31 32]
  [15 16 17 33 34 35]]]
Shape of the resultant array : (2, 3, 6)
```

Summary

If we concatenate any nd array, the result will be same as that of the input array dimension.

```
1-D + 1-D = 1-D
2-D + 2-D = 2-D
3-D + 3-D = 3-D
```

Concatenation is always based on the existing axis. All input arrays must be in the same dimensions. In 2-D arrays (after concatenation).

axis-0 → Number of 2-D arrays are increased and rows and columns remain unchanged.

axis-1 → Number of rows will be increased and the 2-D arrays and columns remain unchanged.

axis-2 → Number of columns will be increased and the 2-D arrays and rows remain unchanged.

Note:

- It is not possible to concatenate arrays with shapes (3,2,3) and (2,1,3) on any axis.

- But axis=none is possible. In this case, both arrays will be flattened to 1-D and then concatenation of arrays can be performed.

8.3   stack()

All input arrays must have the same shape. The resultant stacked array has one more dimension than the input arrays. The joining is always based on a new axis of the newly created array.

```
1-D + 1-D = 2-D
2-D + 2-D = 3-D
```

Stacking 1-D array

To use stack() method, make sure all input arrays must have same shapes, otherwise an error will occur.

Example

```
import numpy as np9
a = np9.array([10,20,30])
b = np9.array([40,50,60,70])
print(np9.stack((a,b)))
```

OUTPUT

```
C:\h_numpy>py h9.py
Traceback (most recent call last):
  File "C:\h_numpy\h9.py", line 4, in <module>
    print(np9.stack((a,b)))
  File "<__array_function__ internals>", line 5, in stack
  File"C:\Users\SureshS\AppData\Local\Programs\Python\Python39\lib\
  site- packages\numpy\core\ shape_base.py", line 426, in stack
  raise ValueError('all inp9ut arrays must have the same shape')
ValueError: all inp9ut arrays must have the same shape
```

Stacking between 1-D arrays

The resultant array will have one more dimension, i.e., 2-D array. The newly created array is 2-D and it has two axes, i.e., axis-0 and axis-1. So we can perform stacking on axis-0 and axis-1.

Stacking along axis-0 in 1-D array

if axis-0, the elements of input array are stacked row-wise. Read row-wise from input arrays and arrange row-wise in the resultant array.

Program 12: A program to print resultant array after stacking using axis=0.

```
import numpy as np9
a = np9.array([10,20,30])
b = np9.array([40,50,60])
resultant_array = np9.stack((a,b)) # default axis=0
print("Resultant array : \n ",resultant_array)
print("Resultant array shape: ",resultant_array.shape)
OUTPUT
C:\h_numpy>py h9.py
Resultant array :
 [[10 20 30]
  [40 50 60]]
Resultant array shape: (2, 3)
```

Stacking along axis-1 in 1-D array

If axis-1, the elements of input array are stacked column-wise. Read row-wise from input arrays and arrange column-wise in the resultant array.

Program 13: A program to print resultant array after stacking using axis=1.

```
import numpy as np9
a = np9.array([10,20,30])
b = np9.array([40,50,60])
resultant_array=np9.stack((a,b),axis=1)
print("Resultant array : \n ",resultant_array)
print("Resultant array shape: ",resultant_array.shape)
OUTPUT
C:\h_numpy>py h9.py
Resultant array :
 [[10 40]
  [20 50]
  [30 60]]
Resultant array shape: (3, 2)
```

Stacking 2-D array

The resultant array will be 3-D array.

$$3\text{-D array shape: } (x, y, z)$$

x → axis-0 ----> number of 2-D arrays
y → axis-1 ----> number of rows in every 2-D array
z → axis-2 ----> number of columns in every 2-D array
axis-0 means 2-D arrays one by one
axis-1 means row wise in each 2-D array
axis-2 means column wise in each 2-D array

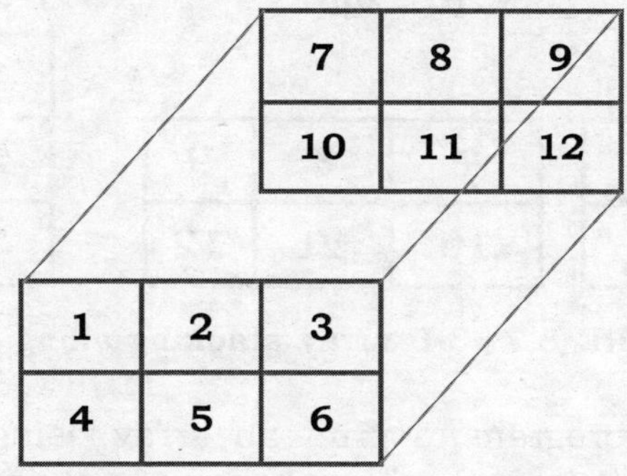

axis-0 means 2-D arrays one by one

FIGURE 8.6   3-D array stacking using axis-0

Program 14: A program to print an array using stacking of 2-D arrays with axis-1 (axis-1 means row wise in each 2-D array)

```
import numpy as np9
a = np9.array([[1,2,3],[4,5,6]])
b = np9.array([[7,8,9],[10,11,12]])
print(np9.stack((a,b),axis=1))
```

OUTPUT

```
C:\h_numpy>py h9.py
 [[[ 1  2  3]
   [ 7  8  9]]

  [[ 4  5  6]
   [10  11  12]]]
```

axis-2 means column wise in each 2-D array

1. **Take first row from array a(1,2,3) and make it as first column in the resultant array**

2. **Take first row from array b(4,5,6) and make it as second column in the resultant array**

3. **Combine these two columns to form a 2-D array**

4. **Repeat steps 1,2,3 for remaining rows of the arrays to make columns**

first 2-D array

[[1 7],

[2 8],

[3 9]]

Second 2-D array

[[4 10],

[5 11],

[6 12]]

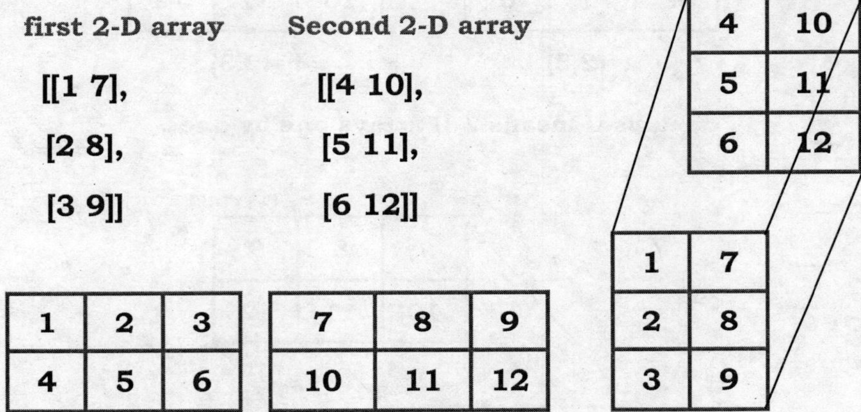

FIGURE 8.7 3-D array stacking along axis-2

Program 15: A program to print an array using stacking of 2-D arrays with axis-2 (axis-2 means column wise in each 2-D array).

```
import numpy as np9
a = np9.array([[1,2,3],[4,5,6]])
b = np9.array([[7,8,9],[10,11,12]])
print(np9.stack((a,b),axis=2))
```

OUTPUT

```
C:\h_numpy>py h9.py
[[[ 1  7]
  [ 2  8]
  [ 3  9]]

 [[ 4 10]
  [ 5 11]
  [ 6 12]]]
```

Program 16: A program to print an array using stacking of three 2-D arrays.

```
import numpy as np9
a = np9.arange(1,7).reshape(3,2)
```

```
b = np9.arange(7,13).reshape(3,2)
c = np9.arange(13,19).reshape(3,2)
print("array a :\n ",a)
print("array b :\n ",b)
print("array c :\n ",c)
# stacking along axis-0
# In 3-D array axis-0 means the number of 2-d arrays
axis0_stack = np9.stack((a,b,c),axis=0)
print("Stacking three 2-D arrays along axis-0:\n ",axis0_stack)
```

OUTPUT

```
C:\h_numpy>py h9.py
array a :
 [[1 2]
  [3 4]
  [5 6]]
array b :
 [[ 7  8]
  [ 9 10]
  [11 12]]
array c :
 [[13 14]
  [15 16]
  [17 18]]
Stacking three 2-D arrays along axis-0:
 [[[ 1  2]
   [ 3  4]
   [ 5  6]]

  [[ 7  8]
   [ 9 10]
   [11 12]]

  [[13 14]
   [15 16]
   [17 18]]]
```

Program 17: A program to print an array using stacking along axis-2 (in 3-D array axis-2 means the number of columns in every 2-D array. Stacking column wise).

```
import numpy as np9
a = np9.arange(1,7).reshape(3,2)
b = np9.arange(7,13).reshape(3,2)
c = np9.arange(13,19).reshape(3,2)
axis2_stack = np9.stack((a,b,c),axis=2)
print(f"Stacking three 2-D arrays along axis-2:\n ",axis2_stack)
```

OUTPUT

```
C:\h_numpy>py h9.py
Stacking three 2-D arrays along axis-2:
```

```
[[[ 1  7 13]
  [ 2  8 14]]

 [[ 3  9 15]
  [ 4 10 16]]

 [[ 5 11 17]
  [ 6 12 18]]]
```

Note:

- Reading of arrays row-wise
- Arranging is based on the newly created array axis

Program 18: A program to print an array after stacking three 1-D arrays.

```python
import numpy as np9
a = np9.arange(4)
b = np9.arange(4,8)
c = np9.arange(8,12)
print("array a :",a)
print("array b :",b)
print("array c :",c)
# We will get 2-D array
# In 2-D array avaialble axes are: axis-0 and axis-1
# Based on axis-0:
# axis-0 in 2-D array means the number of rows
axis0_stack = np9.stack((a,b,c),axis=0)
print("Stacking three 2-D arrays along axis-0:\n ",axis0_stack)
# Based on axis-1:
# axis-1 in 2-D array means the number of columns
axis1_stack = np9.stack((a,b,c),axis=1)
print("Stacking three 2-D arrays along axis-1:\n ",axis1_stack)
```

OUTPUT

```
C:\h_numpy>py h9.py
array a : [0 1 2 3]
array b : [4 5 6 7]
array c : [ 8 9 10 11]
Stacking three 2-D arrays along axis-0:
 [[ 0  1  2  3]
  [ 4  5  6  7]
  [ 8  9 10 11]]
Stacking three 2-D arrays along axis-1:
 [[ 0  4  8]
  [ 1  5  9]
  [ 2  6 10]
  [ 3  7 11]]
```

Difference between concatenate() and stack()

Table 8.1 concatenate() vs stack()

concatenate()	stack()
It joins a sequence of arrays along an existing axis.	Here joining is based on a new axis.
The dimension of newly created array is same as that of input array.	The dimension of newly created array is more than the input array dimension.
To perform concatenation, all input arrays must have same dimensions. The size of all dimensions except concatenation axis must be the same.	To perform stack operation, it is mandatory that all input arrays must have same shapes, i.e., dimensions and sizes should be the same.

8.4 vstack()

vstack stands for the vertical stack. It is used for joining arrays based on axis-0. For 1-D arrays, we get a 2-D array as output. For the remaining dimensions, it acts as concatenate () along axis-0.

Rules

- The input arrays must have the same shapes except the first axis (axis-0).
- 1-D arrays must have the same size.
- The array formed by stacking the given arrays will be at least 2-D.
- vstack() operation is equivalent to concatenation along the first axis when 1-D arrays of shape (N,) is reshaped to (1, N).
- For 2-D or more dimension arrays, vstack() simply acts as concatenation w.r.t. axis-0.

Program 19: A program to print an array using vstack() for 1-D arrays. (vstack for 1-D arrays of same sizes)

```
import numpy as np9
a = np9.array([10,20,30,40])
b = np9.array([50,60,70,80])
# a will be converted to shapes (1,4) and b will be converted
to (1,4)
print(np9.vstack((a,b)))
```

OUTPUT

```
C:\h_numpy>py h9.py
[[10 20 30 40]
 [50 60 70 80]]
```

For 2-D arrays

The sizes of axis-1 should be same to perform vstack() with concatenation rule.

Program 20: A program to print an array using vstack() for 2-D arrays [Here it is possible because sizes of axis-1 are same (3 and 3). vstack() is always performed along axis-0].

```
import numpy as np9
a = np9.arange(1,10).reshape(3,3)
b = np9.arange(10,16).reshape(2,3)
print(np9.vstack((a,b)))
```
OUTPUT
```
C:\h_numpy>py h9.py
[[ 1  2  3]
 [ 4  5  6]
 [ 7  8  9]
 [10 11 12]
 [13 14 15]]
```

Program 21: A program to print an array using vstack() for 2-D arrays. [Here it is not possible because sizes of axis-1 are different (3 and 2). vstack() is always performed along axis-0].

```
import numpy as np9
a = np9.arange(1,10).reshape(3,3)
b = np9.arange(10,16).reshape(3,2)
print(np9.vstack((a,b)))
```

OUTPUT

```
C:\h_numpy>py h9.py
Traceback (most recent call last):
  File "C:\h_numpy\h9.py", line 7, in <module>
    print(np9.vstack((a,b)))
  File "<__array_function__ internals>", line 5, in vstack
  File "C:\Users\Suresh S\AppData\Local\Programs\Python\Python39\
  lib\ site-packages\numpy\core\shape_base.py", line 282, in
  vstack
    return _nx.concatenate(arrs, 0)
  File "<__array_function__ internals>", line 5, in concatenate
```
ValueError: all the inp9ut array dimensions for the concatenation axis must match exactly, but along dimension 1, the array at index 0 has size 3 and the array at index 1 has size 2

For 3-D arrays

axis-0 means the number of 2-D arrays.

Program 22: A program to print an array using vstack() for 3-D arrays.

```
import numpy as np9
a = np9.arange(1,25).reshape(2,3,4)
b = np9.arange(25,49).reshape(2,3,4)
print("array a : \n ",a)
```

```
print("array b : \n ",b)
result = np9.vstack((a,b))
print("Result of vstack : \n ",result)
```

OUTPUT

```
C:\h_numpy>py h9.py
array a :
 [[[ 1  2  3  4]
  [ 5  6  7  8]
  [ 9 10 11 12]]

 [[13 14 15 16]
  [17 18 19 20]
  [21 22 23 24]]]
array b :
 [[[25 26 27 28]
  [29 30 31 32]
  [33 34 35 36]]

 [[37 38 39 40]
  [41 42 43 44]
  [45 46 47 48]]]
Result of vstack :
 [[[ 1  2  3  4]
  [ 5  6  7  8]
  [ 9 10 11 12]]

 [[13 14 15 16]
  [17 18 19 20]
  [21 22 23 24]]

 [[25 26 27 28]
  [29 30 31 32]
  [33 34 35 36]]

 [[37 38 39 40]
  [41 42 43 44]
  [45 46 47 48]]]
```

8.5 hstack()

hstack stands for horizontal stack. It is same as concatenate() but joining is always based on axis-1.

$$1\text{-}D + 1\text{-}D \rightarrow 1\text{-}D$$

Rules

- This is equivalent to concatenation along the second axis, except for 1-D arrays where it concatenates along the first axis.

- All input arrays must be of same dimensions.
- Except axis-1, all the remaining sizes must be equal.

Program 23: A program to print an array using hstack() for 1-D arrays.

```
import numpy as np9
a = np9.array([10,20,30,40])
b = np9.array([50,60,70,80,90,100])
print(np9.hstack((a,b)))
```

OUTPUT

```
C:\h_numpy>py h9.py
[ 10 20 30 40 50 60 70 80 90 100]
```

Program 24: A program to print an array using hstack() for 2-D arrays.

```
import numpy as np9
a = np9.arange(1,7).reshape(3,2)
b = np9.arange(7,16).reshape(3,3)
print(np9.hstack((a,b)))
```

OUTPUT

```
C:\h_numpy>py h9.py
[[ 1  2  7  8  9]
 [ 3  4 10 11 12]
 [ 5  6 13 14 15]]
```

8.6   dstack()

dstack() stands for depth/height stack. It is used for concatenation based on axis-2. 1-D and 2-D arrays will be converted to a 3-D arrays. The resultant should be at least 3-D array.

Rules

- This is equivalent to concatenation along the third axis after 2-D arrays of shape (M,N) is reshaped to (M,N,1) and 1-D arrays of shape (N,) is reshaped to (1,N,1).
- The arrays must have the same shapes except the third axis.
- 1-D or 2-D arrays must have the same shapes.
- The array formed by stacking the given arrays will be at least 3-D.

Program 25: A program to print an array using dstack() for 1-D arrays.

```
import numpy as np9
a = np9.array([1,2,3])
b = np9.array([2,3,4])
print(np9.dstack((a,b)))
```

OUTPUT

```
C:\h_numpy>py h9.py
[[[1 2]
  [2 3]
  [3 4]]]
```

Program 26: A program to print an array using dstack() for 1-D arrays.

```
import numpy as np9
a = np9.array([[1],[2],[3]])
b = np9.array([[2],[3],[4]])
print(np9.dstack((a,b)))
```

OUTPUT

```
C:\h_numpy>py h9.py
[[[1 2]]
  [[2 3]]
  [[3 4]]]
```

Summary of Joining of nd Arrays

Table 8.2   Summary of joining of nd arrays

concatenate()	Join a sequence of arrays along an existing axis.
stack()	Join a sequence of arrays along a new axis.
vstack()	Stack arrays in sequence vertically according to first axis (axis-0).
hstack()	Stack arrays in sequence horizontally according to second axis (axis-1).
dstack()	Stack arrays in sequence depth wise according to third axis (axis-2).

8.7   SPLITTING OF ARRAYS

We can perform split operation on nd arrays using the following functions:

1. split()
2. vsplit()
3. hsplit()
4. dsplit()
5. array_split()

We will get only views, but not copies because the data is not going to be changed in split().

Syntax

```
split(array, indices_or_sections, axis=0)
```

Note:

- It splits an array into multiple sub-arrays of equal size.
- Sections mean the number of sub-arrays. It returns a list of nd array objects.
- All sections must be of equal sizes, otherwise an error occurs.

split() based on sections

- We can split arrays based on sections or indices.
- If we split arrays based on sections, the sizes of sub-arrays should be equal.
- If we split arrays based on indices, then the sizes of sub-arrays need not to be the same.

Program 27: A program to print an array using split () for 1-D arrays (axis-0).

```
import numpy as np9
a = np9.arange(1,10)
sub_arrays = np9.split(a,3)
print("array a : ",a)
print("Type of sub_arrays :",type(sub_arrays))
print("sub_arrays : ",sub_arrays)
```

OUTPUT

```
C:\h_numpy>py h9.py
array a : [1 2 3 4 5 6 7 8 9]
Type of sub_arrays : <class 'list'>
sub_arrays : [array([1, 2, 3]), array([4, 5, 6]), array([7, 8, 9])]
```

Note:

- If dividing array into equal number of specified sections is not possible, then we will get error.

Example

```
import numpy as np9
a = np9.arange(1,10)
sub_arrays = np9.split(a,4)
print("array a : ",a)
print("Type of sub_arrays :",type(sub_arrays))
print("sub_arrays : ",sub_arrays)
```

OUTPUT

```
C:\h_numpy>py h9.py
Traceback (most recent call last):
  File "C:\h_numpy\h9.py", line 3, in <module>
    sub_arrays = np9.split(a,4)
  File "<__array_function__ internals>", line 5, in split
  File "C:\Users\Suresh S\AppData\Local\Programs\Python\Python39\
  lib\ site-packages\numpy\lib\shape_base.py", line 872, in split
  raise ValueError(

ValueError: array split does not result in an equal division
```

2-D arrays (axis-0 ==> Vertical split)

Splitting is based on axis-0 by default, i.e., row-wise split (vertical split). We can also split based on axis-1, i.e., column-wise split (horizontal split).

Program 28: A program for splitting based on default axis, i.e., axis-0 (vertical split).

```
import numpy as np9
a = np9.arange(1,25).reshape(6,4)
result_3sections = np9.split(a,3) # dividing 3 sections vertically
print("array a : \n ",a)
print(f"splitting the array into 3 sections along axis-0 : \n ",result_3sections)
```

Note: Here we can use various possible sections: 2, 3, 6.

OUTPUT

```
C:\h_numpy>py h9.py
array a :
 [[ 1  2  3  4]
 [ 5  6  7  8]
 [ 9 10 11 12]
 [13 14 15 16]
 [17 18 19 20]
 [21 22 23 24]]
```

splitting the array into 3 sections along axis-0 :
```
[array([[1, 2, 3, 4],
[5, 6, 7, 8]]), array([[ 9, 10, 11, 12],
[13, 14, 15, 16]]), array([[17, 18, 19, 20],
[21, 22, 23, 24]])]
```

Note: Here we can use various possible sections: 2, 3, 6.

```
import numpy as np9
a = np9.arange(1,25).reshape(6,4)
result_2sections = np9.split(a,2,axis=0)
result_6sections = np9.split(a,6,axis=0)
print("splitting the array into 2 sections along axis-0 : \n ",result_2sections)
print("splitting the array into 6 sections along axis-0 : \n ",result_6sections)
```

OUTPUT

```
C:\h_numpy>py h9.py
splitting the array into 2 sections along axis-0 :
[array([[ 1,  2,  3,  4],
       [ 5,  6,  7,  8],
       [ 9, 10, 11, 12]]),
 array([[13, 14, 15, 16],
       [17, 18, 19, 20],
       [21, 22, 23, 24]])]
```

splitting the array into 6 sections along axis-0 :
[array([[1, 2, 3, 4]]), array([[5, 6, 7, 8]]), array([[9, 10, 11, 12]]), array([[13, 14, 15, 16]]), array([[17, 18, 19, 20]]), array([[21, 22, 23, 24]])]

2-D arrays (axis-1 ==> Horizontal split)

Program 29: A program for splitting based on axis-1 (horizontal split).

for the shape(6,4) we can perform 4 or 2 sections

```
import numpy as np9
a = np9.arange(1,25).reshape(6,4)
result_2sections = np9.split(a,2,axis=1) # dividing 2 sections horizontally
result_4sections = np9.split(a,4,axis=1) # dividing 4 sections horizontally
print("splitting the array into 2 sections along axis-0 : \n ",result_2sections)
print("splitting the array into 4 sections along axis-0 : \n ",result_4sections)
```

OUTPUT

```
C:\h_numpy>py h9.py
splitting the array into 2 sections along axis-0 :
[array([[ 1,  2],
 [ 5,  6],
 [ 9, 10],
 [13, 14],
 [17, 18],
 [21, 22]]), array([[ 3,  4],
 [ 7,  8],
 [11, 12],
 [15, 16],
 [19, 20],
 [23, 24]])]
splitting the array into 4 sections along axis-0 :
[array([[ 1],
 [ 5],
 [ 9],
 [13],
 [21]]), array([[ 2],
 [ 6],
 [10],
 [14],
 [18],
 [22]]), array([[ 3],
 [ 7],
 [11],
 [15],
 [19],
```

```
[23]]]), array([[ 4],
[ 8],
[12],
[16],
[20],
[24]])]
```

split() based on indices

We can also split arrays based on indices. The sizes of sub-arrays need not to be equal.

1-D arrays (axis-0)

array([10,20,30,40,50,60,70,80,90,100])   **np9.split(a,[3,7])**

0	1	2	3	4	5	6	7	8	9
10	20	30	40	50	60	70	80	90	100

defalt axis-0
[3,7] are indices
total 3 sub arrays are created
sub array-1 --> before index 3 -->0,1,2
sub array-2 --> from index 3 to before index 7 -->3,4,5,6
sub array-3 --> from index 7 to last index -->7,8,9

FIGURE 8.8   split() based on indices of 1-D array

Program 30: A program for splitting the 1-D array based on indices.

```
import numpy as np9
a = np9.arange(10,101,10)
result = np9.split(a,[3,7])
print("array a : ",a)
print("splitting the 1-D array based on indices : \n ",result)
```

OUTPUT

```
C:\h_numpy>py h9.py
array a : [ 10 20 30 40 50 60 70 80 90 100]
splitting the 1-D array based on indices :
 [array([10, 20, 30]), array([40, 50, 60, 70]), array([ 80, 90,
100])]
```

Program 31: A program for splitting the 1-D array based on indices.

```
import numpy as np9
a = np9.arange(10,101,10)
result = np9.split(a,[2,5,7])
# [2,5,7] ==> 4 subarrays
# subarray-1 : before index-2 ==> 0,1
# subarray-2 : from index-2 to before index-5 ==> 2,3,4
# subarray-3 : from index-5 to before index-7 ==> 5,6
# subarray-4 : from index-7 to last index ==> 7,8,9
print("array a : ",a)
print("splitting the 1-D array based on indices : \n ",result)
```

OUTPUT

```
C:\h_numpy>py h9.py
array a : [ 10 20 30 40 50 60 70 80 90 100]
```

```
splitting the 1-D array based on indices :
[array([10, 20]), array([30, 40, 50]), array([60, 70]), array([
80, 90, 100])]
```

2-D arrays (axis-0)

split based on indices and axis-0
Vertical split

		np9.split(a,[3,4]) or
array ([[1,2],	**0**	**np9.split(a,[3,4], axis-0)**
[3.4],	**1**	total 3 sub arrays
		sub array-1 --> before index 3 -->0,1,2
[5,6],	**2**	sub array-2 --> from index 3 to before
[7,8],	**3**	index 4 -->3
[9,10],	**4**	sub array-3 --> from index 4 to last index
		-->4,5
[11,12])	**5**	

FIGURE 8.9   split() based on indices of 2-D array

Program 32: A program for splitting 2-D arrays based on indices along axis=0.

```python
import numpy as np9
a = np9.arange(1,13).reshape(6,2)
result = np9.split(a,[3,4])
print("array a : \n ",a)
print("resultant array after vertical split : \n ",result)
```

OUTPUT

```
C:\h_numpy>py h9.py
array a :
 [[ 1  2]
 [ 3  4]
 [ 5  6]
 [ 7  8]
 [ 9  10]
 [11  12]]
resultant array after vertical split :
 [array([[1,  2],
[3,  4],
[5,  6]]), array([[7,  8]]), array([[ 9,  10],
[11,  12]])]
```

2-D arrays (axis-1)

split based on indices and axis-1 Horizontal split   **np9.split(a,[1,3,5], axis-1)**

array ([[1, 2, 3, 4, 5, 6],	For [1,3,5] 4 sub arrays are created
	sub array-1 --> before index 1 -->0
[7, 8, 9, 10, 11, 12],	sub array-2 --> from index 1 to before
	index 3 -->1,2
[13, 14, 15, 16, 17, 18]])	sub array-3 --> from index 3 to before
	index 5 -->3,4
	sub array-4 --> from index 5 to last index
	5 -->5

FIGURE 8.10   split() on 1-D array for axis=1

Example

```
import numpy as np9
a = np9.arange(1,19).reshape(3,6)
result = np9.split(a,[1,3,5],axis=1)
print(f"array a : \n ",a)
print(f"resultant array after horizontal split : \n ",result)
```

OUTPUT

```
C:\h_numpy>py h9.py
array a :
 [[ 1  2  3  4  5  6]
  [ 7  8  9 10 11 12]
  [13 14 15 16 17 18]]
resultant array after horizontal split :
[array([[ 1],
 [ 7],
 [13]]), array([[ 2,  3],
 [ 8,  9],
 [14, 15]]), array([[ 4,  5],
 [10, 11],
 [16, 17]]), array([[ 6],
 [12],
 [18]])]
```

Example

```
import numpy as np9
a = np9.arange(1,19).reshape(3,6)
result = np9.split(a,[2,4,4],axis=1)
# [2,4,4] => 4 subarrays are created
# subarray1: 2 ==> before index-2 ==> 0,1
# subarray2: 4 ==> from index-2 to before index-4 ==> 2,3
# subarray3: 4 ==> from index-4 to before index-4 ==> empty array
# subarray4: ==> from index-4 to last index ==> 4,5
print("array a : \n ",a)
print("resultant array after horizontal split : \n ",result)
print("first subarray : \n ",result[0])
print("second subarray : \n ",result[1])
print("third subarray : \n ",result[2])
print("fourth subarray : \n ",result[3])
```

OUTPUT

```
C:\h_numpy>py h9.py
array a :
 [[ 1  2  3  4  5  6]
  [ 7  8  9 10 11 12]
  [13 14 15 16 17 18]]
resultant array after horizontal split :
 [array([[ 1,  2],
  [ 7,  8],
  [13, 14]]), array([[ 3,  4],
  [ 9, 10],
```

```
[15, 16]]), array([], shape=(3, 0), dtype=int32), array([[ 5, 6],
[11, 12],
[17, 18]])]
first subarray :
 [[ 1  2]
 [ 7  8]
 [13 14]]
second subarray :
 [[ 3  4]
 [ 9 10]
 [15 16]]
third subarray :
 []
fourth subarray :
 [[ 5  6]
 [11 12]
 [17 18]]
```

Program 33: A program to perform splitting of an array into sub arrays.

```python
import numpy as np9
a = np9.arange(1,19).reshape(3,6)
result = np9.split(a,[0,2,6],axis=1)
# [0,2,6] => 4 subarrays are created
# subarray1: 0 ==> before index-0 ==> empty
# subarray2: 2 ==> from index-0 to before index-2 ==> 0,1
# subarray3: 6 ==> from index-2 to before index-6 ==> 2,3,4,5
# subarray4: ==> from index-6 to last index ==> empty
print("array a : \n ",a)
print("resultant array after horizontal split : \n ",result)
print("first subarray : \n ",result[0])
print("second subarray : \n ",result[1])
print("third subarray : \n ",result[2])
print("fourth subarray : \n ",result[3])
```

OUTPUT

```
C:\h_numpy>py h9.py
array a :
 [[ 1  2  3  4  5  6]
 [ 7  8  9 10 11 12]
 [13 14 15 16 17 18]]
resultant array after horizontal split :
 [array([], shape=(3, 0), dtype=int32), array([[ 1, 2],
 [ 7, 8],
 [13, 14]]), array([[ 3, 4, 5, 6],
 [ 9, 10, 11, 12],
 [15, 16, 17, 18]]), array([], shape=(3, 0), dtype=int32)]
first subarray :
 []
second subarray :
 [[ 1  2]
```

```
 [ 7  8]
 [13 14]]
third subarray :
 [[ 3  4  5  6]
 [ 9 10 11 12]
 [15 16 17 18]]
fourth subarray :
 []
```

Example

```python
import numpy as np9
a = np9.arange(1,19).reshape(3,6)
result = np9.split(a,[1,5,3],axis=1)
# [1,5,3] => 4 subarrays are created
# subarray1: 1 ==> before index-1 ==> 0
# subarray2: 5 ==> from index-1 to before index-5 ==> 1,2,3,4
# subarray3: 3 ==> from index-5 to before index-3 ==> empty
# subarray4: ==> from index-3 to last index ==> 3,4,5
print("array a : \n ",a)
print("resultant array after horizontal split : \n ",result)
print("first subarray : \n ",result[0])
print("second subarray : \n ",result[1])
print("third subarray : \n ",result[2])
print("fourth subarray : \n ",result[3])
```

OUTPUT

```
C:\h_numpy>py h9.py
array a :
 [[ 1  2  3  4  5  6]
 [ 7  8  9 10 11 12]
 [13 14 15 16 17 18]]
resultant array after horizontal split :
[array([[ 1],
 [ 7],
 [13]]), array([[ 2,  3,  4,  5],
 [ 8,  9, 10, 11],
 [14, 15, 16, 17]]), array([], shape=(3, 0), dtype=int32),
 array([[ 4,  5,  6],
 [10, 11, 12],
 [16, 17, 18]])]
first subarray :
 [[ 1]
 [ 7]
 [13]]
second subarray :
 [[ 2  3  4  5]
 [ 8  9 10 11]
 [14 15 16 17]]
third subarray :
 []
```

```
fourth subarray :
 [[ 4  5  6]
  [10 11 12]
  [16 17 18]]
```

8.8 vsplit()

The vsplit() function is used to split an array vertically (row wise). The split is based on axis-0.

1-D arrays

To use vsplit, input array should be atleast 2-D array. It is not possible to split 1-D array vertically.

Example

```
import numpy as np9
a = np9.arange(10)
print(np9.vsplit(a,2))
```

OUTPUT

```
C:\h_numpy>py h9.py
Traceback (most recent call last):
  File "C:\h_numpy\h9.py", line 3, in <module>
   print(np9.vsplit(a,2))
  File "<__array_function__ internals>", line 5, in vsplit
  File "C:\Users\Suresh S\AppData\Local\Programs\Python\Python39\
  lib\ site-packages\numpy\lib\shape_base.py", line 990, in vsplit
  raise ValueError('vsplit only works on arrays of 2 or more
  dimensions')
```

```
ValueError: vsplit only works on arrays of 2 or more dimensions
```

2-D arrays

Example

```
import numpy as np9
a = np9.arange(1,13).reshape(6,2)
print(np9.vsplit(a,2))
```

OUTPUT

```
C:\h_numpy>py h9.py
[array([[1,  2],
 [3,  4],
 [5,  6]]), array([[ 7,  8],
 [ 9, 10],
 [11, 12]])]
```

Example

```
import numpy as np9
a = np9.arange(1,13).reshape(6,2)
print(np9.vsplit(a,3))
```

OUTPUT

```
C:\h_numpy>py h9.py
[array([[1, 2],
 [3, 4]]), array([[5, 6],
 [7, 8]]), array([[ 9, 10],
 [11, 12]])]
```

Example

```
import numpy as np9
a = np9.arange(1,13).reshape(6,2)
print(np9.vsplit(a,6))
```

OUTPUT

```
C:\h_numpy>py h9.py
[array([[1, 2]]), array([[3, 4]]), array([[5, 6]]),
array([[7, 8]]), array([[ 9, 10]]), array([[11, 12]])]
```

8.9   hsplit()

The hsplit() function is used to split an array horizontally (column wise). The split is based on second axis (axis-1).

Example

```
import numpy as np9
a = np9.arange(10)
print(np9.hsplit(a,2))
```

OUTPUT

```
C:\h_numpy>py h9.py
[array([0, 1, 2, 3, 4]), array([5, 6, 7, 8, 9])]
```

2-D arrays

Based on axis-1 only

Example

```
import numpy as np9
a = np9.arange(1,13).reshape(3,4)
print(np9.hsplit(a,2))
```

OUTPUT

```
C:\h_numpy>py h9.py
[array([[ 1, 2],
 [ 5, 6],
 [ 9, 10]]), array([[ 3, 4],
 [ 7, 8],
 [11, 12]])]
```

Program 34: A program for splitting an array using hsplit() based on indices.

```
import numpy as np9
a = np9.arange(10,101,10)
print(np9.hsplit(a,[2,4]))
```

OUTPUT

```
C:\h_numpy>py h9.py
[array([10, 20]), array([30, 40]), array([ 50, 60, 70, 80, 90,
100])]
```

Program 35: A program for splitting an array using hsplit() based on indices.

hsplit() based on indices:

```
import numpy as np9
a = np9.arange(24).reshape(4,6)
print(np9.hsplit(a,[2,4]))
```

OUTPUT

```
C:\h_numpy>py h9.py
[array([[ 0,  1],
 [ 6,  7],
 [12,  13],
 [18,  19]]), array([[ 2,  3],
 [ 8,  9],
 [14,  15],
 [20,  21]]), array([[ 4,  5],
 [10,  11],
 [16,  17],
 [22,  23]])]
```

8.10   dsplit()

dsplit stands for depth split. It is used for splitting based on third axis (axis-2).

Example

```
import numpy as np9
a = np9.arange(24).reshape(2,3,4)
print(a)
```

OUTPUT

```
C:\h_numpy>py h9.py
[[[ 0  1  2  3]
  [ 4  5  6  7]
  [ 8  9  10  11]]

 [[12 13 14 15]
  [16 17 18 19]
  [20 21 22 23]]]
```

Program 36: A program for splitting an array using dsplit based on sections.

```
import numpy as np9
a = np9.arange(24).reshape(2,3,4)
print(np9.dsplit(a,2))
```

OUTPUT

```
C:\h_numpy>py h9.py
[array([[[ 0,  1],
  [ 4,  5],
  [ 8,  9]],
 [[12,  13],
  [16,  17],
  [20,  21]]]), array([[[ 2,  3],
  [ 6,  7],
  [10,  11]],
 [[14,  15],
  [18,  19],
  [22,  23]]])]
```

Program 37: A program for splitting array with dsplit based on indices.

```
import numpy as np9
a = np9.arange(24).reshape(2,3,4)
print(np9.dsplit(a,[1,3]))
```

OUTPUT

```
C:\h_numpy>py h9.py
[array([[[ 0],
  [ 4],
  [ 8]],

 [[12],
  [16],
  [20]]]), array([[[ 1,  2],
  [ 5,  6],
  [ 9,  10]],

 [[13,  14],
  [17,  18],
  [21,  22]]]), array([[[ 3],
  [ 7],
  [11]],

 [[15],
  [19],
  [23]]])]
```

8.11 Array_split()

In the case of split() with sections, the array should be split into equal parts. If equal parts are not possible, then we will get an error. But in the case of array_split(), we won't get any error. The only difference between split() and array_split() is that 'array_split' allows 'indices_or_ sections' to be an integer that does not equally divide the axis.

For an array of length x that should be split into n sections, it returns x % n sub-arrays of size x//n + 1 and the rest of size x//n.

Example

```
#x % n sub-arrays of size x//n + 1
# and the rest of size x//n
# 10 elements --->3 sections
# 10%3(1) sub-arrays of size 10//3+1(4)
# and the rest(2) of size 10//3 (3)
# 1 sub-array of size 4 and the rest of size 3
#(4,3,3)
import numpy as np9
a = np9.arange(10,101,10)
print(np9.array_split(a,3))
```

OUTPUT

```
C:\h_numpy>py h9.py
[array([10, 20, 30, 40]), array([50, 60, 70]), array([ 80, 90,
100])]
```

Example

```
# 11 elements 3 sections
# it returns x % n (11%3=2)sub-arrays of size x//n + 1(11//3+1=4)
# and the rest(1) of size x//n.(11//3=3)
# (4,4,3)
import numpy as np9
a = np9.arange(11)
print(np9.array_split(a,3))
```

OUTPUT

```
C:\h_numpy>py h9.py
[array([0, 1, 2, 3]), array([4, 5, 6, 7]), array([ 8, 9, 10])]
```

Example

```
# 2-D array
# x=6 n=4
# x % n sub-arrays of size x//n + 1-->2 sub-arrays of size:2
# rest of size x//n.--->2 sub-arrays of size:1
# 2,2,1,1,
import numpy as np9
a = np9.arange(24).reshape(6,4)
print(a)
```

OUTPUT

```
C:\h_numpy>py h9.py
 [[ 0  1  2  3]
  [ 4  5  6  7]
  [ 8  9 10 11]
  [12 13 14 15]
  [16 17 18 19]
  [20 21 22 23]]
```

Example

```
import numpy as np9
a = np9.arange(24).reshape(6,4)
print(np9.array_split(a,4))
```

OUTPUT

```
C:\h_numpy>py h9.py
[array([[0, 1, 2, 3],
    [4, 5, 6, 7]]), array([[ 8, 9, 10, 11],
    [12, 13, 14, 15]]), array([[16, 17, 18, 19]]), array([[20, 21,
    22, 23]])]
```

Summary of Split Methods

Table 8.3 Summary of split methods

split()	Split an array into multiple sub-arrays of equal size. Raise error if an equal division cannot be made.
vsplit()	Split an array into multiple sub-arrays vertically (row wise).
hsplit()	Split an array into multiple sub-arrays horizontally (column wise).
dsplit()	Split an array into multiple sub-arrays along the third axis (depth).
array_split()	Split an array into multiple sub-arrays of equal or near-equal size. Does not raise an exception if an equal division cannot be made.

EXERCISES

1. Write a program to check that if the shape of result and out differs then we will get error: ValueError.

2. Write a program to print an array after concatenation of two 2-D arrays.

3. Write a program to print an array after concatenation along axis-1.

4. Write a program to print an array using stacking of 2-D arrays with axis-0 (axis-0 means 2-D arrays one by one).

5. Write a program to print an array using stacking along axis-1 (In 3-D array, axis-1 means the number of rows. Stacking row wise).

6. Write a program to print an array using vstack() for 1-D arrays. (vstack for 1-D arrays of different sizes).

7. Write a program for splitting an array using vsplit() based on indices.

9   SORTING AND SEARCHING ELEMENTS OF nd ARRAY

9.1   SORTING ELEMENTS OF nd ARRAYS

We can sort elements of nd array. NumPy module contains sort() function. The default sorting algorithm is quicksort and it is in the ascending order. We can also specify mergesort, heapsort, etc. for numbers in the ascending order and for strings in the alphabetical order.

Program 1: A program for sorting 1-D arrays.

```
import numpy as np9
a = np9.array([70,20,60,10,50,40,30])
sorted_array = np9.sort(a)
print("Original array a : ",a)
print("Sorted array(Ascending by default : ",sorted_array)
```

OUTPUT

```
C:\h_numpy>py h9.py
Original array a : [70 20 60 10 50 40 30]
Sorted array(Ascending by default : [10 20 30 40 50 60 70]
```

Program 2: A program for sorting the 1-D arrays in the descending order.

1st way :: np9.sort(a)[::-1]

```
import numpy as np9
a = np9.array([70,20,60,10,50,40,30])
sorted_array = np9.sort(a)[::-1]
print(f"Original array a : ",a)
print(f"Sorted array(Descending) : ",sorted_array)
```

OUTPUT

```
C:\h_numpy>py h9.py
Original array a : [70 20 60 10 50 40 30]
Sorted array(Descending) : [70 60 50 40 30 20 10]
```

2nd way :: -np9.sort(-a)

```
import numpy as np9
a = np9.array([70,20,60,10,50,40,30])
print(-a)
```

OUTPUT

```
C:\h_numpy>py h9.py
[-70 -20 -60 -10 -50 -40 -30]
```

2nd way :: -np9.sort(-a)

```
import numpy as np9
a = np9.array([70,20,60,10,50,40,30])
print(np9.sort(-a)) # Ascending order
```

OUTPUT

```
C:\h_numpy>py h9.py
[-70 -60 -50 -40 -30 -20 -10]
```

Descending the 1-D arrays
2nd way :: -np9.sort(-a)

```
import numpy as np9
a = np9.array([70,20,60,10,50,40,30])
print(-np9.sort(-a)) # Descending Order
```

OUTPUT

```
C:\h_numpy>py h9.py
[70 60 50 40 30 20 10]
```

2-D arrays

- axis-0 --->the number of rows (axis= –2)
- axis-1--->the number of columns (axis= –1) ==> default value
- Sorting is based on columns in 2-D arrays by default. Every 1-D array will be sorted.

Example

```
import numpy as np9
a= np9.array([[40,20,70],[30,20,60],[70,90,80]])
ascending = np9.sort(a)
print("Original array a :\n ",a)
print("Sorted array(Ascending) : \n ",ascending)
```

OUTPUT

```
C:\h_numpy>py h9.py
Original array a :
 [[40 20 70]
 [30 20 60]
 [70 90 80]]
Sorted array(Ascending) :
 [[20 40 70]
 [20 30 60]
 [70 80 90]]
```

9.2 order PARAMETER

Use the order keyword to specify a field to sort a structured array.

Program 3: A program for creating the structured array.

```
import numpy as np9
dtype = [('name', 'S10'), ('height', float), ('age', int)]
values = [('Gopie', 1.7, 45), ('Vikranth', 1.5, 38),('Sathwik',
1.8, 28)]
a = np9.array(values, dtype=dtype)
sort_height = np9.sort(a, order='height')
sort_age = np9.sort(a,order='age')
print("Original Array :\n ",a)
print("Sorting based on height :\n ",sort_height)
print("Sorting based on age :\n ",sort_age)
```

OUTPUT

```
C:\h_numpy>py h9.py
Original Array :
 [(b'Gopie', 1.7, 45) (b'Vikranth', 1.5, 38) (b'Sathwik', 1.8, 28)]
Sorting based on height :
 [(b'Vikranth', 1.5, 38) (b'Gopie', 1.7, 45) (b'Sathwik', 1.8, 28)]
Sorting based on age :
 [(b'Sathwik', 1.8, 28) (b'Vikranth', 1.5, 38) (b'Gopie', 1.7, 45)]
```

9.3 SEARCHING ELEMENTS OF nd ARRAYS

We can search elements of nd array by using where() function.

Syntax

```
where(condition, [x, y])
```

- If we specify the condition only, then returns the indices of the elements which satisfy the condition.
- If we provide condition, x, y then the elements which satisfy the condition will be replaced with x and the remaining elements will be replaced with y.
- where() function does not return the elements. It returns only the indices.
- It will act as replacement operator also.
- The functionality is just similar to ternary operator when it acts as replacement operator.

Program 4: A program to find indexes where the value is 7 from 1-D array.

```
import numpy as np9
a = np9.array([3,5,7,6,7,9,4,6,10,15])
b = np9.where(a==7)
print(b) # element 7 is available at 2 and 4 indices
```

OUTPUT
```
C:\h_numpy>py h9.py
(array([2, 4], dtype=int64),)
```

9.4 FINDING THE ELEMENTS DIRECTLY

We can get the elements directly in the following two ways:

1. using where() function
2. using condition-based selection

Program 5: A program to get the odd numbers using where() function.

```
import numpy as np9
a = np9.array([3,5,7,6,7,9,4,6,10,15])
indices = np9.where(a%2!=0)
print(a[indices])
```

OUTPUT
```
C:\h_numpy>py h9.py
[ 3  5  7  7  9 15]
```

Program 6: A program to get the odd numbers using condition based selection.

```
import numpy as np9
a = np9.array([3,5,7,6,7,9,4,6,10,15])
print(a[a%2!=0])
```

OUTPUT
```
C:\h_numpy>py h9.py
[ 3  5  7  7  9 15]
```

Program 7: A program to check the condition "where (condition, [x, y])", if condition is satisfied then the elements are replaced by x and if the condition fails then the elements are replaced by y. Replace every even number with 8888 and every odd number with 7777.

```
import numpy as np9
a = np9.array([3,5,7,6,7,9,4,6,10,15])
b = np9.where( a%2 == 0, 8888, 7777)
print(b)
```

OUTPUT
```
C:\h_numpy>py h9.py
[7777 7777 7777 8888 7777 7777 8888 8888 8888 7777]
```

We can use where() function for any n-dimensional array. In 2-D arrays,

- It returns two arrays.
- First array is row indices.
- Second array is column indices.

Program 8: A program to find the indices of the elements where elements are divisible by 5.

```
import numpy as np9
a = np9.arange(12).reshape(4,3)
print(np9.where(a%5==0))
```

OUTPUT

```
C:\h_numpy>py h9.py
(array([0, 1, 3], dtype=int64), array([0, 2, 1], dtype=int64))
```

Note:

- The first array, array([0, 1, 3] represents the row indices.
- The second array, array([0, 2, 1] represents the column indices.
- The required elements are present at (0, 0), (1, 2) and (3, 1) index places.

Program 9: A program to perform replacement on 2-D arrays.

```
import numpy as np9
a = np9.arange(12).reshape(4,3)
print(np9.where(a%5==0,9999,a))
```

OUTPUT

```
C:\h_numpy>py h9.py
[[9999 1    2]
 [ 3    4 9999]
 [ 6    7    8]
 [ 9 9999 11]]
```

9.5   searchsorted() FUNCTION

Internally this function uses the binary search algorithm. Hence we can call this function only for sorted arrays. If the array is not sorted then we will get abnormal results. The complexity of the binary search algorithm is O (log n). It returns the insertion point (i.e., index) of the given element.

Program 10: A program to find the insertion point of 6 from left.

```
import numpy as np9
a = np9.arange(0,31,5)
print(np9.searchsorted(a,6))
```

OUTPUT

```
C:\h_numpy>py h9.py
2
```

Note:

- By default, it will always search from left-hand side to identify insertion points.
- If we want to search from right-hand side, we should use side='right'.

Program 11: A program to find the insertion point of 6 from right.

```
import numpy as np9
a = np9.arange(0,31,5)
print("Array a :\n ",a)
np9.searchsorted(a,6,side='right')
print(np9.searchsorted(a,6))
```

OUTPUT

```
C:\h_numpy>py h9.py
Array a :
 [ 0 5 10 15 20 25 30]
2
```

Program 12: A program to find the insertion point from left and right.

```
import numpy as np9
a = np9.array([3,5,7,6,7,9,4,10,15,6])
# first sort the elements
a = np9.sort(a)
# insertion point from left(default)
left = np9.searchsorted(a,6)
# insertion point from right
right = np9.searchsorted(a,6,side='right')
print("The original array : ",a)
print("Insertion point for 6 from left : ",left)
print("Insertion point for 6 from right : ",right)
```

OUTPUT

```
C:\h_numpy>py h9.py
The original array : [ 3 4 5 6 6 7 7 9 10 15]
Insertion point for 6 from left : 3
Insertion point for 6 from right : 5
```

Summary

Table 9.1   Summary of elements in nd array

sort()	To sort given array.
where()	To perform search and replace operation.
searchsorted()	To identify insertion point in the given sorted array.

EXERCISES

1. Write a program to sort string elements in the alphabetical order.

2. Write a program to find indices where odd numbers are present in the given 1-D array.

3. Write a program to find indexes where odd numbers are present in the given 1-D array and replace it with element 9999.

10 INSERT AND DELETE ELEMENTS OF nd ARRAY

10.1   INSERT ELEMENTS INTO nd ARRAY

To insert an element into an array, there are following two functions:
    insert()   inserting the element at the required position
    append()   inserting the element at the end

10.2   insert() FUNCTION

Insert values along the given axis before the given indices.

Syntax

```
insert(array, obj, values, axis=None)
```
where
        obj-->Object that defines the index or indices before which
        'values' are inserted.
        values--->Values to insert into array.
        axis ---->Axis along which to insert 'values'.

Program 1: A program to insert 7777 before index 2.

```
import numpy as np9
a = np9.arange(10)
b = np9.insert(a,2,7777)
print("array a : ",a)
print("array b : ",b)
```

OUTPUT

```
C:\h_numpy>py h9.py
array a : [0 1 2 3 4 5 6 7 8 9]
array b : [ 0 1 7777 2 3 4 5 6 7 8 9]
```

Program 2: A program to insert 7777 before indexes 2 and 5.

```
import numpy as np9
a = np9.arange(10)
b = np9.insert(a,[2,5],7777)
print("array a : ",a)
print("array b : ",b)
```

OUTPUT

```
C:\h_numpy>py h9.py
array a : [0 1 2 3 4 5 6 7 8 9]
array b : [ 0 1 7777 2 3 4 7777 5 6 7 8 9]
```

CASE STUDY ON SHAPE MISMATCH

```
import numpy as np9
a = np9.arange(10)
b = np9.insert(a,[2,5],[7777,8888,9999])
print("array a : ",a)
print("array b : ",b)
```

OUTPUT

```
C:\h_numpy>py h9.py
Traceback (most recent call last):
  File "C:\h_numpy\h9.py", line 4, in <module>
   b = np9.insert(a,[2,5],[7777,8888,9999])
  File "<__array_function__ internals>", line 5, in insert
  File"C:\Users\Suresh
  S\AppData\Local\Programs\Python\Python39\lib\site-packages\numpy\
  lib\function_base.py", line 4750, in insert
  new[tuple(slobj)] = values
ValueError: shape mismatch: value array of shape (3,) could not
be broadcast to indexing result of shape (2,)
```

Another way

```
import numpy as np9
a = np9.arange(10)
b = np9.insert(a,[2,5,7],[7777,8888])
print("array a : ",a)
print("array b : ",b)
```

OUTPUT

```
C:\h_numpy>py h9.py
Traceback (most recent call last):
  File "C:\h_numpy\h9.py", line 4, in <module>
   b = np9.insert(a,[2,5,7],[7777,8888])
  File "<__array_function__ internals>", line 5, in insert
  File"C:\Users\Suresh
  S\AppData\Local\Programs\Python\Python39\lib\site-packages\numpy\
  lib\ function_base.py", line 4750, in insert
  new[tuple(slobj)] = values
ValueError: shape mismatch: value array of shape (2,) could not
be broadcast  to indexing result of shape (3,)
```

Way 2

```
import numpy as np9
a = np9.arange(10)
b = np9.insert(a,[2,5,5],[777,888,999])
print("array a : ",a)
print("array b : ",b)
```

OUTPUT

```
C:\h_numpy>py h9.py
array a : [0 1 2 3 4 5 6 7 8 9]
array b : [ 0 1 777 2 3 4 888 999 5 6 7 8 9]
```

Program 3: A program to check IndexError.

```
import numpy as np9
a = np9.arange(10)
b = np9.insert(a,25,7777)
print("array a : ",a)
print("array b : ",b)
```

OUTPUT

```
C:\h_numpy>py h9.py
Traceback (most recent call last):
  File "C:\h_numpy\h9.py", line 4, in <module>
   b = np9.insert(a,25,7777)
  File "<__array_function__ internals>", line 5, in insert
  File "C:\Users\Suresh
  S\AppData\Local\Programs\Python\Python39\lib\site-packages\numpy\
  lib\ function_base.py", line 4704, in insert
    raise IndexError(
IndexError: index 25 is out of bounds for axis 0 with size 10
```

Note:

- All the insertion points (indices) are identified at the beginning of the insert operation.
- The array should contain only homogeneous elements.
- By using the insert() function, if we are trying to insert any other type of the element, then that element will be converted to array type automatically before insertion.
- If the conversion is not possible then we will get an error.

Example

```
import numpy as np9
a = np9.arange(10)
b = np9.insert(a,2,True)
print("array a : ",a)
print("array b : ",b)
```

OUTPUT

```
C:\h_numpy>py h9.py
array a : [0 1 2 3 4 5 6 7 8 9]
array b : [0 1 1 2 3 4 5 6 7 8 9]
```

Program 4: A program with NumPy and string.

```
import numpy as np9
a = np9.arange(10)
b = np9.insert(a,2,'GK')
print("array a : ",a)
print("array b : ",b)
```

OUTPUT

```
C:\h_numpy>py h9.py
Traceback (most recent call last):
  File "C:\h_numpy\h9.py", line 4, in <module>
   b = np9.insert(a,2,'GK')
  File "<__array_function__ internals>", line 5, in insert
  File "C:\Users\Suresh S\AppData\Local\Programs\Python\Python39\
  lib\ site-packages\numpy\lib\ function_base.py", line 4712, in
  insert
  values = array(values, copy=False, ndmin=arr.ndim, dtype=arr.
  dtype)
ValueError: invalid literal for int() with base 10: 'GK'
```

Program 5: A program to print complex number.

```
import numpy as np9
a = np9.arange(10)
b = np9.insert(a,2,10+20j)
print("array a : ",a)
print("array b : ",b)
```

OUTPUT

```
C:\h_numpy>py h9.py
Traceback (most recent call last):
  File "C:\h_numpy\h9.py", line 4, in <module>
   b = np9.insert(a,2,10+20j)
  File "<__array_function__ internals>", line 5, in insert
  File "C:\Users\Suresh S\AppData\Local\Programs\Python\Python39\
  lib\site-packages\numpy\lib\function_base.py", line 4712, in
  insert
  values = array(values, copy=False, ndmin=arr.ndim, dtype=arr.
  dtype)
TypeError: can't convert complex to int
```

Summary for 1-D arrays while insertion

- The number of indices and the number of elements should be matched.
- Out of range index is not allowed.
- Elements will be converted automatically to the array type.

2-D arrays

If we are trying to insert elements into multi-dimensional arrays, it is essential to provide an axis value. If the axis value is not provided, then the default value (none) will be considered. In this case, the array flattens to a 1-D array and then insertion can be performed.

axis=0 means rows (axis= –2)

axis=1 means columns (axis= –1)

Program 6: A program to check if the axis is not defined for 2-D arrays. None is selected by default (here the 2-D array flattens to 1-D array and insertion will be performed).

```
import numpy as np9
a = np9.array([[10,20],[30,40]])
b = np9.insert(a,1,100)
print("array a :\n ",a)
print("array b : ",b)
```

OUTPUT

```
C:\h_numpy>py h9.py
array a :
 [[10 20]
  [30 40]]
array b : [ 10 100 20 30 40]
```

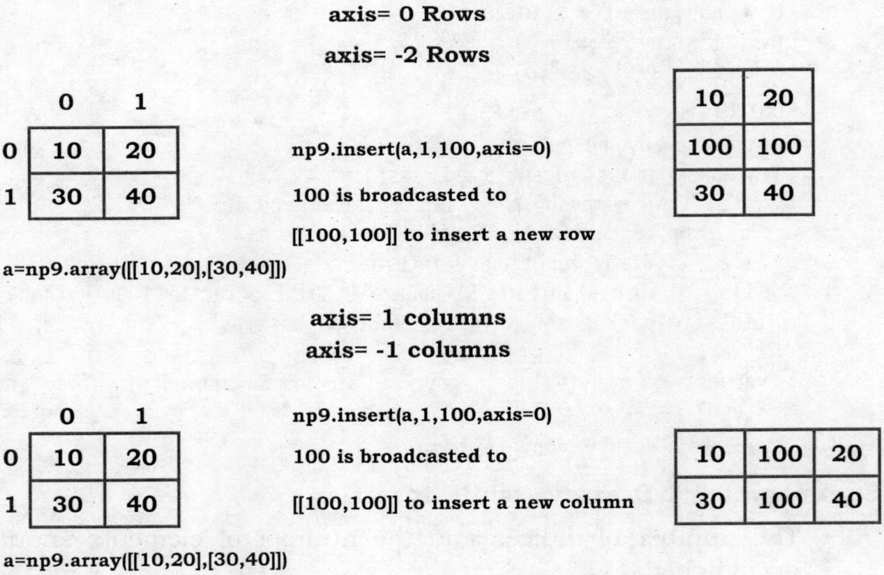

FIGURE 10.1   2-D array flatten to 1-D array

Program 7: A program to insert the elements along axis=1 or axis= –1 (rows are inserted).

```
import numpy as np9
a = np9.array([[10,20],[30,40]])
b = np9.insert(a,1,100,axis=1)
c = np9.insert(a,1,100,axis=-1)
```

```
print("array a :\n ",a)
print("array b :\n ",b)
print("array c :\n ",c)
```

OUTPUT

```
C:\h_numpy>py h9.py
array a :
 [[10 20]
  [30 40]]
array b :
 [[ 10 100  20]
  [ 30 100  40]]
array c :
 [[ 10 100  20]
  [ 30 100  40]]
```

Program 8: A program to insert multiple rows.

```
import numpy as np9
a = np9.array([[10,20],[30,40]])
b = np9.insert(a,1,[[100,200],[300,400]],axis=0)
print("array a :\n ",a)
print("array b :\n ",b)
```

OUTPUT

```
C:\h_numpy>py h9.py
array a :
 [[10 20]
  [30 40]]
array b :
 [[ 10  20]
  [100 200]
  [300 400]
  [ 30  40]]
```

Program 9: A program to insert multiple columns.

```
import numpy as np9
a = np9.array([[10,20],[30,40]])
b = np9.insert(a,1,[[100,200],[300,400]],axis=1)
print("array a :\n ",a)
print("array b :\n ",b)
```

OUTPUT

```
C:\h_numpy>py h9.py
array a :
 [[10 20]
  [30 40]]
array b :
 [[ 10 100 300  20]
  [ 30 200 400  40]]
```

CASE STUDY

```
# ValueError
        import numpy as np9
        a = np9.array([[10,20],[30,40]])
        b = np9.insert(a,1,[100,200,300],axis=0)
        print("array a :\n ",a)
        print("array b :\n ",b)
```

OUTPUT

```
C:\h_numpy>py h9.py
Traceback (most recent call last):
  File "C:\h_numpy\h9.py", line 4, in <module>
   b = np9.insert(a,1,[100,200,300],axis=0)
  File "<__array_function__ internals>", line 5, in insert
  File "C:\Users\Suresh S\AppData\Local\Programs\Python\Python39\
  lib\ site-packages\numpy\lib\function_base.py", line 4724, in
  insert
  new[tuple(slobj)] = values
ValueError: could not broadcast inp9ut array from shape (1,3)
into shape (1,2)
```

10.3 append()

By using the insert() function, we can insert elements at our required index position. If we want to add elements at the end of the nd array, then we have to use the append() function.

Program 10: A program to append 100 elements to the existing array of int.

```
import numpy as np9
a = np9.arange(10)
b = np9.append(a,100)
print("array a : ",a)
print("array b : ",b)
```

OUTPUT

```
C:\h_numpy>py h9.py
array a : [0 1 2 3 4 5 6 7 8 9]
array b : [ 0 1 2 3 4 5 6 7 8 9 100]
array b : [ 0. 1. 2. 3. 4. 5. 6. 7. 8. 9. 20.5]
```

Program 11: A program to append string at index.

```
import numpy as np9
a = np9.arange(10)
b = np9.append(a,"GK")
print("array a : ",a)
print("array b : ",b)
```

OUTPUT

```
C:\h_numpy>py h9.py
array a : [0 1 2 3 4 5 6 7 8 9]
array b : ['0' '1' '2' '3' '4' '5' '6' '7' '8' '9' 'GK']
```

Program 12: A program to append Boolean value at index.

```
import numpy as np9
a = np9.arange(10)
b = np9.append(a,True)
print("array a : ",a)
print("array b : ",b)
```

OUTPUT

```
C:\h_numpy>py h9.py
array a : [0 1 2 3 4 5 6 7 8 9]
array b : [0 1 2 3 4 5 6 7 8 9 1]
```

Program 13: A program to append complex value at index.

```
import numpy as np9
a = np9.arange(10)
b = np9.append(a,20+15j)
print("array a : ",a)
print("array b : ",b)
```

OUTPUT

```
C:\h_numpy>py h9.py
array a : [0 1 2 3 4 5 6 7 8 9]
array b : [ 0. +0.j 1. +0.j 2. +0.j 3. +0.j 4. +0.j 5. +0.j
6. +0.j 7. +0.j 8. +0.j    9. +0.j 20.+15.j]
```

2-D arrays

If we are not specifying the axis, then the input array will be
flattened to the 1-D array, and then append will be performed. If we
are providing an axis, then all the input arrays must have a same
number of dimensions, and the same shape as the provided axis.
If axis=0 is taken then the obj should match with the number of
columns of the input array. If axis=1 is taken then the obj should
match with the number of rows of the input array.

Program 14: A program where the axis is not specified so the
default value "None" will be taken and flatten to 1-D array and
then the insertion will be performed.

```
import numpy as np9
a = np9.array([[10,20],[30,40]])
b = np9.append(a,70)
print("array a :\n ",a)
print("array b : ",b)
```

OUTPUT

```
C:\h_numpy>py h9.py
array a :
 [[10 20]
  [30 40]]
array b : [10 20 30 40 70]
```

CASE STUDY

```
# axis=0 ==> along rows
# Value Error : i/p array is 2-D and appended array is 0-D
```

```
import numpy as np9
a = np9.array([[10,20],[30,40]])
b = np9.append(a,70,axis=0) # appended is 0-D => 70
print("array a :\n ",a)
print("array b : ",b)
```

OUTPUT

```
C:\h_numpy>py h9.py
Traceback (most recent call last):
  File "C:\h_numpy\h9.py", line 5, in <module>
    b = np9.append(a,70,axis=0) # appended is 0-D => 70
  File "<__array_function__ internals>", line 5, in append
  File "C:\Users\Suresh S\AppData\Local\Programs\Python\Python39\
  lib\ site-packages\numpy\lib\function_base.py", line 4817, in
  append
    return concatenate((arr, values), axis=axis)
  File "<__array_function__ internals>", line 5, in concatenate
ValueError: all the inp9ut arrays must have same number of
dimensions, but the array at index 0 has 2 dimension(s) and the
array at index 1 has 0 dimension(s)
```

CASE STUDY

```
# Value Error : i/p array is 2-D and appended array is 1-D
import numpy as np9
a = np9.array([[10,20],[30,40]])
b = np9.append(a,[70,80],axis=0) # appended is 1-D => [70,80]
print("array a :\n ",a)
print("array b : ",b)
```

OUTPUT

```
C:\h_numpy>py h9.py
Traceback (most recent call last):
  File "C:\h_numpy\h9.py", line 4, in <module>
    b = np9.append(a,[70,80],axis=0) # appended is 1-D => [70,80]
  File "<__array_function__ internals>", line 5, in append
  File "C:\Users\Suresh S\AppData\Local\Programs\Python\Python39\
  lib\ site-packages\numpy\lib\function_base.py", line 4817, in
  append
    return concatenate((arr, values), axis=axis)
  File "<__array_function__ internals>", line 5, in concatenate
ValueError: all the inp9ut arrays must have same number of
dimensions, but the array at index 0 has 2 dimension(s) and the
array at index 1 has 1 dimension(s)
```

Example: # i/p array is 2-D and appended array is 2-D.

```
import numpy as np9
a = np9.array([[10,20],[30,40]])
b = np9.append(a,[[70,80]],axis=0) # appended is 2-D => [[70,80]]
print("array a :\n ",a)
print("array b : ",b)
```

OUTPUT

```
C:\h_numpy>py h9.py
array a :
 [[10  20]
  [30  40]]
array b :  [[10  20]
  [30  40]
  [70  80]]
```

CASE STUDY

axis=1 along columns
Value Error : i/p array is size 2 and appended element is size 1

```
import numpy as np9
a = np9.array([[10,20],[30,40]])
b = np9.append(a,[[70,80]],axis=1)  # appended is size 1 =>
[[70,80]]
print("array a :\n ",a)
print("array b : ",b)
```

OUTPUT

```
C:\h_numpy>py h9.py
Traceback (most recent call last):
  File "C:\h_numpy\h9.py", line 5, in <module>
   b = np9.append(a,[[70,80]],axis=1) # appended is size 1 => [[70,80]]
  File "<__array_function__ internals>", line 5, in append
  File "C:\Users\Suresh S\AppData\Local\Programs\Python\Python39\
  lib\ site-packages\numpy\lib\function_base.py", line 4817, in
  append
   return concatenate((arr, values), axis=axis)
  File "<__array_function__ internals>", line 5, in concatenate
ValueError: all the inp9ut array dimensions for the concatenation
axis must match exactly, but along dimension 0, the array at
index 0 has size 2 and the array at index 1 has size 1
```

Example

```
import numpy as np9
a = np9.array([[10,20],[30,40]])
b = np9.append(a,[[70],[80]],axis=1)
print("array a :\n ",a)
print("array b : ",b)
```

OUTPUT

```
C:\h_numpy>py h9.py
array a :
 [[10  20]
  [30  40]]
array b :
 [[10  20  70]
  [30  40  80]]
```

Program 15: A program to append multiple columns.

```
import numpy as np9
a = np9.array([[10,20],[30,40]])
b = np9.append(a,[[70,80],[90,100]],axis=1)
print("array a :\n ",a)
print("array b : ",b)
OUTPUT
C:\h_numpy>py h9.py
array a :
 [[10 20]
  [30 40]]
array b :
 [[10 20 70]
  [30 40 80]]
C:\h_numpy>py h9.py
array a :
 [[10 20]
  [30 40]]
array b :
 [[ 10 20 70 80]
  [ 30 40 90 100]]
```

CASE STUDY

```
# Consider the array Which of the following operations will be
performed successfully?
# A. np9.append(a,[[10,20,30]],axis=0) #valid
# B. np9.append(a,[[10,20,30]],axis=1) #invalid
# C. np9.append(a,[[10],[20],[30]],axis=0) #invalid
# D. np9.append(a,[[10],[20],[30],[40]],axis=1) #valid
# E. np9.append(a,[[10,20,30],[40,50,60]],axis=0) #valid
# F. np9.append(a,[[10,20],[30,40],[50,60],[70,80]],axis=1) #valid
import numpy as np9
a = np9.arange(12).reshape(4,3)
print(np9.append(a,[[10,20,30]],axis=0)) #valid
```

OUTPUT

```
C:\h_numpy>py h9.py
 [[ 0  1  2]
  [ 3  4  5]
  [ 6  7  8]
  [ 9 10 11]
  [10 20 30]]
import numpy as np9
a = np9.arange(12).reshape(4,3)
print(np9.append(a,[[10,20,30]],axis=1)) #invalid
```

OUTPUT

```
C:\h_numpy>py h9.py
Traceback (most recent call last):
  File "C:\h_numpy\h9.py", line 3, in <module>
    print(np9.append(a,[[10,20,30]],axis=1)) #invalid
```

```
  File "<__array_function__ internals>", line 5, in append
  File "C:\Users\Suresh S\AppData\Local\Programs\Python\Python39\
  lib\ site-packages\numpy\lib\function_base.py", line 4817, in
  append
  return concatenate((arr, values), axis=axis)
  File "<__array_function__ internals>", line 5, in concatenate
ValueError: all the inp9ut array dimensions for the concatenation
axis must match exactly, but along dimension 0, the array at
index 0 has size 4 and the array at index 1 has size 1
```

```python
import numpy as np9
a = np9.arange(12).reshape(4,3)
print(np9.append(a,[[10],[20],[30]],axis=0)) #invalid
import numpy as np9
a = np9.arange(12).reshape(4,3)
print(np9.append(a,[[10],[20],[30],[40]],axis=1)) #valid
```

OUTPUT

```
C:\h_numpy>py h9.py
 [[ 0  1  2 10]
  [ 3  4  5 20]
  [ 6  7  8 30]
  [ 9 10 11 40]]
import numpy as np9
a = np9.arange(12).reshape(4,3)
print(np9.append(a,[[10,20,30],[40,50,60]],axis=0)) #valid
```

OUTPUT

```
C:\h_numpy>py h9.py
 [[ 0  1  2]
  [ 3  4  5]
  [ 6  7  8]
  [ 9 10 11]
  [10 20 30]
  [40 50 60]]
import numpy as np9
a = np9.arange(12).reshape(4,3)
print(np9.append(a,[[10,20],[30,40],[50,60],[70,80]],axis=1))
#valid
```

OUTPUT

```
:\h_numpy>py h9.py
[[ 0  1  2 10 20]
 [ 3  4  5 30 40]
 [ 6  7  8 50 60]
 [ 9 10 11 70 80]]
```

Example I

```python
import numpy as np9
a = np9.arange(12).reshape(4,3)
print(a)
b = np9.append(a,[[10],[20],[30],[40]],axis=1)
print(b)
```

OUTPUT

```
C:\h_numpy>py h9.py
[[ 0  1  2]
 [ 3  4  5]
 [ 6  7  8]
 [ 9 10 11]]

[[ 0  1  2 10]
 [ 3  4  5 20]
 [ 6  7  8 30]
 [ 9 10 11 40]]
```

Example II

```
import numpy as np9
a = np9.arange(12).reshape(4,3)
print(a)
b = np9.append(a,[[10,50],[20,60],[30,70],[40,80]],axis=1)
print(b)
```

OUTPUT

```
C:\h_numpy>py h9.py
 [[ 0   1  2]
  [ 3   4  5]
  [ 6   7  8]
  [ 9  10 11]]

 [[ 0  1  2 10 50]
  [ 3  4  5 20 60]
  [ 6  7  8 30 70]
  [ 9 10 11 40 80]]
```

insert() vs append()

- By using insert() function, we can insert elements at our required index position.
- By using append() function, we can add elements at the end of nd array.

10.4 DELETE ELEMENTS FROM nd ARRAY

We can delete elements of nd array by using delete() function.

Syntax

```
delete(arr, obj, axis=None)
```

Here

- obj can be int, an array of ints, or slice.
- For multi-dimensional arrays, we must specify the axis, otherwise the default axis=none will be considered. In this case, first the array is flattened to the 1-D array and then deletion will be performed.

1-D arrays

Program 16: A program to delete a single element of 1-D array at a specified index.

```
import numpy as np9
a = np9.arange(10,101,10)
b = np9.delete(a,3) # to delete the element present at 3rd index
print("array a : ",a)
print("array b : ",b)
```

OUTPUT

```
C:\h_numpy>py h9.py
array a : [ 10 20 30 40 50 60 70 80 90 100]
array b : [ 10 20 30 50 60 70 80 90 100]
```

Program 17: A program to delete elements of 1-D array at the specified indices.

```
import numpy as np9
a = np9.arange(10,101,10)
b = np9.delete(a,[0,4,6]) # to delete elements present at
indices:0,4,6
print("array a : ",a)
print("array b : ",b)
```

OUTPUT

```
C:\h_numpy>py h9.py
array a : [ 10 20 30 40 50 60 70 80 90 100]
array b : [ 20 30 40 60 80 90 100]
```

2-D arrays

We have to provide axis. If the axis is not specified, the array will be flattened (1-D) and then deletion will be performed.

Program 18: A program where axis is not provided and the default axis=none will be taken.

```
import numpy as np9
a = np9.arange(1,13).reshape(4,3)
b = np9.delete(a,1)
print("array a : ",a)
print("array b : ",b)
```

OUTPUT

```
C:\h_numpy>py h9.py
array a : [[ 1 2 3]
 [ 4 5 6]
 [ 7 8 9]
 [10 11 12]]
array b : [ 1 3 4 5 6 7 8 9 10 11 12]
```

Program 19: A program in which if axis=0, it deletes the specified row.

```
import numpy as np9
a = np9.arange(1,13).reshape(4,3)
b = np9.delete(a,1,axis=0) # row at index 1 will be deleted
```

```
print("array a : ",a)
print("array b : ",b)
```

OUTPUT

```
C:\h_numpy>py h9.py
array a : [[ 1  2  3]
 [ 4  5  6]
 [ 7  8  9]
 [10 11 12]]
array b : [[ 1  2  3]
 [ 7  8  9]
 [10 11 12]]
```

Program 20: A program where axis=0 and it deletes the specified row.

```
import numpy as np9
a = np9.arange(1,13).reshape(4,3)
b = np9.delete(a,[1,3],axis=0) # row at index 1 and index 3
will be deleted
print("array a : ",a)
print("array b : ",b)
```

OUTPUT

```
C:\h_numpy>py h9.py
array a : [[ 1  2  3]
 [ 4  5  6]
 [ 7  8  9]
 [10 11 12]]
array b : [[1 2 3]
 [7 8 9]]
```

Program 21: A program where axis=0 and it deletes the specified range of rows.

```
import numpy as np9
a = np9.arange(1,13).reshape(4,3)
b = np9.delete(a,np9.s_[0:3],axis=0) # rows from index-0 to
index-(3-1) will be deleted
print("array a : ",a)
print("array b : ",b)
```

OUTPUT

```
C:\h_numpy>py h9.py
array a : [[ 1  2  3]
 [ 4  5  6]
 [ 7  8  9]
 [10 11 12]]
array b : [[10 11 12]]
```

Program 22: A program in which if axis=0, it deletes the specified range of rows with step value.

```
import numpy as np9
a = np9.arange(1,13).reshape(4,3)
b = np9.delete(a,np9.s_[::2],axis=0) # to delete every 2nd row
```

```
(alternative row) from index-0
print("array a : ",a)
print("array b : ",b)
```

OUTPUT

```
C:\h_numpy>py h9.py
array a : [[ 1  2  3]
 [ 4  5  6]
 [ 7  8  9]
 [10 11 12]]
array b : [[ 4 5 6]
 [10 11 12]]
```

Program 23: A program where axis=1 and it deletes the specified column.

```
import numpy as np9
a = np9.arange(1,13).reshape(4,3)
b = np9.delete(a,1,axis=1) # column at index 1 will be deleted
print("array a : ",a)
print("array b : ",b)
```

OUTPUT

```
C:\h_numpy>py h9.py
array a : [[ 1  2  3]
 [ 4  5  6]
 [ 7  8  9]
 [10 11 12]]
array b : [[ 1 3]
 [ 4  6]
 [ 7  9]
 [10 12]]
```

Program 24: A program in which if axis=1, it deletes the specified column.

```
import numpy as np9
a = np9.arange(1,13).reshape(4,3)
b = np9.delete(a,[1,2],axis=1) # columns at index 1 and index
2 will be deleted
print("array a : ",a)
print("array b : ",b)
```

OUTPUT

```
C:\h_numpy>py h9.py
array a : [[ 1  2  3]
 [ 4  5  6]
 [ 7  8  9]
 [10 11 12]]
array b : [[ 1]
 [ 4]
 [ 7]
 [10]]
```

Program 25: A program where if axis=1, it deletes the specified range of columns.

```python
import numpy as np9
a = np9.arange(1,13).reshape(4,3)
b = np9.delete(a,np9.s_[0:2],axis=1) # rows from index-0 to
index-(2-1) will be deleted
print("array a : ",a)
print("array b : ",b)
```

OUTPUT

```
C:\h_numpy>py h9.py
array a : [[ 1  2  3]
 [ 4  5  6]
 [ 7  8  9]
 [10 11 12]]
array b : [[ 3]
 [ 6]
 [ 9]
 [12]]
```

3-D arrays

Program 26: A program where axis=none.

```python
import numpy as np9
a = np9.arange(24).reshape(2,3,4)
b = np9.delete(a,3) # flatten to 1-D array and element at 3rd
index will be deleted
print("array a : ",a)
print("array b : ",b)
```

OUTPUT

```
C:\h_numpy>py h9.py
array a : [[[ 0  1  2  3]
  [ 4  5  6  7]
  [ 8  9 10 11]]

 [[12 13 14 15]
  [16 17 18 19]
  [20 21 22 23]]]
array b : [ 0  1  2  4  5  6  7  8  9 10 11 12 13 14 15 16 17 18
19 20 21 22 23]
```

Program 27: A program that deletes a 2-D array at the specified index if axis=0.

```python
import numpy as np9
a = np9.arange(24).reshape(2,3,4)
b = np9.delete(a,0,axis=0) # 2-D array at index 0 will be deleted
print("array a : ",a)
print("array b : ",b)
```

OUTPUT

```
C:\h_numpy>py h9.py
```

```
array a : [[[ 0  1  2  3]
  [ 4  5  6  7]
  [ 8  9 10 11]]

 [[12 13 14 15]
  [16 17 18 19]
  [20 21 22 23]]]
array b : [[[12 13 14 15]
  [16 17 18 19]
  [20 21 22 23]]]
```

Program 28: A program that deletes a column at the specified index from every 2-D array, if axis=1.

```
import numpy as np9
a = np9.arange(24).reshape(2,3,4)
b = np9.delete(a,1,axis=2) # column at index 1 will be deleted
from every 2-D array
print("array a : ",a)
print("array b : ",b)
```

OUTPUT

```
C:\h_numpy>py h9.py
array a : [[[ 0  1  2  3]
  [ 4  5  6  7]
  [ 8  9 10 11]]

 [[12 13 14 15]
  [16 17 18 19]
  [20 21 22 23]]]
array b : [[[ 0  2  3]
  [ 4  6  7]
  [ 8 10 11]]

 [[12 14 15]
  [16 18 19]
  [20 22 23]]]
```

CASE STUDY

Consider the following array:
array([[0, 1, 2], [3, 4, 5], [6, 7, 8], [9, 10, 11]])
Delete last row and insert following row in that place: → [70,80,90]

```
# Solution for the case study
# Step:1 ==> create the required ndarray
import numpy as np9
a = np9.arange(12).reshape(4,3)
# Step:2 ==> Delete the last row.
# We have to mention the axis as axis=0 for rows.
# obj value we can take as -1 , which represents the last row
b = np9.delete(a,-1,axis=0)
print("Original Array")
print("array a : \n ",a)
print("After deletion the array ")
```

```
print("array b : \n ",b)
# Step:3 ==> we have to insert the row [70.80.90] at the end
# for this we can use append() function with axis=0
c = np9.append(b,[[70,80,90]],axis=0)
print("After insertion the array ")
print("array c : \n ",c)
```

OUTPUT

```
C:\h_numpy>py h9.py
Original Array
array a :
 [[ 0  1  2]
  [ 3  4  5]
  [ 6  7  8]
  [ 9 10 11]]
After deletion the array
array b :
 [[0  1  2]
  [3  4  5]
  [6  7  8]]
After insertion the array
array c :
 [[ 0  1  2]
  [ 3  4  5]
  [ 6  7  8]
  [70 80 90]]
```

Summary

Table 10.1 Summary of insert, append and delete

insert()	Insert elements into an array at specified index.
append()	Append elements at the end of an array.
delete()	Delete elements from an array.

EXERCISES

1. Write a program to insert 7777 before index 2 and 8888 before index 5.

2. Write a program to check that the original array contains int values. If inserted value is float type, then the float value is converted to the array type, i.e., int. There may be data loss in this scenario.

3. Write a program to insert the elements along axis=0 or axis=-2 (rows are inserted).

4. Write a program to delete elements of 1-D array from a specified range using Python range. This is applicable only for 1-D arrays.

5. Write a program to perform that if axis=1, delete a row at the specified index from every 2-D array.

11 MATRIX MULTIPLICATION USING dot() FUNCTION

11.1 MATRIX CLASS IN NUMPY LIBRARY

A 1-D array is called vector and 2-D array is called matrix. Matrix class is a specially designed class to create 2-D arrays.

Creating 2-D arrays

There are two ways of creating 2-D arrays by using:

1. matrix class
2. nd array class

11.2 MATRIX CLASS (nd ARRAY)

Synatx

```
matrix(data, dtype=None, copy=True)
```

Parameters

1. **data: array_like or string:** If data is a string, it is interpreted as a matrix with commas or spaces separating columns, and semicolons separating rows.

2. **dtype: data type:** Data type of the output matrix.

3. **copy: bool:** If data is already an nd array, then this flag determines whether the data is copied (the default), or whether a view is constructed.

Program 1: A program for creating matrix object from string

```
# a = np9.matrix('col1 col2 col3;col1 col2 col3')
# a = np9.matrix('col1,col2,col3;col1,col2,col3')
import numpy as np9
```

```
a = np9.matrix('10,20;30,40')
b = np9.matrix('10 20;30 40')
print("type of a : type \n",a)
print("type of b : type \n",b)
print("Matrix object creation from string with comma : \n",a)
print("Matrix object creation from string with space : \n",b)
```

OUTPUT

```
C:\h_numpy>py h9.py
type of a : type
 [[10 20]
  [30 40]]
type of b : type
 [[10 20]
  [30 40]]
Matrix object creation from string with comma :
 [[10 20]
  [30 40]]
Matrix object creation from string with space :
 [[10 20]
  [30 40]]
```

Program 2: A program to create a matrix from ndarray.

```
import numpy as np9
a = np9.arange(6).reshape(3,2)
b = np9.matrix(a)
print("type of a : ",type(a) )
print("type of b : :",type(b))
print('ndarray :\n ',a)
print('matrix :\n ',b)
```

OUTPUT

```
C:\h_numpy>py h9.py
type of a : <class 'numpy.ndarray'>
type of b : : <class 'numpy.matrix'>
ndarray :
 [[0 1]
  [2 3]
  [4 5]]
matrix :
 [[0 1]
  [2 3]
  [4 5]]
```

+ operator in nd array and matrix

In case of both nd array and matrix, + operator behaves in the same way.

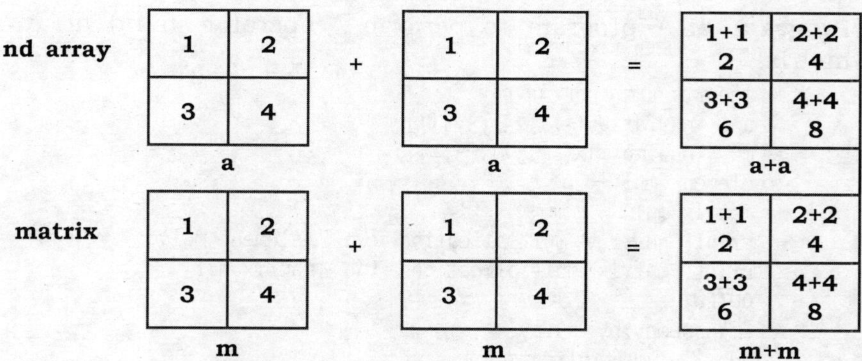

FIGURE 11.1   + operator in nd array

Program 3: A program to perform the + operator in nd array and matrix.

```
import numpy as np9
a = np9.array([[1,2],[3,4]])
m = np9.matrix([[1,2],[3,4]])
addition_a = a+a
addition_m = m+m
print('ndarray addition :\n ',addition_a)
print('matrix addition :\n ',addition_m)
```

OUTPUT

```
C:\h_numpy>py h9.py
ndarray addition :
 [[2 4]
 [6 8]]
matrix addition :
 [[2 4]
 [6 8]]
```

* operator in nd array and matrix

- In case of nd array, * operator performs element level multiplication.
- In case of matrix, * operator performs matrix multiplication.

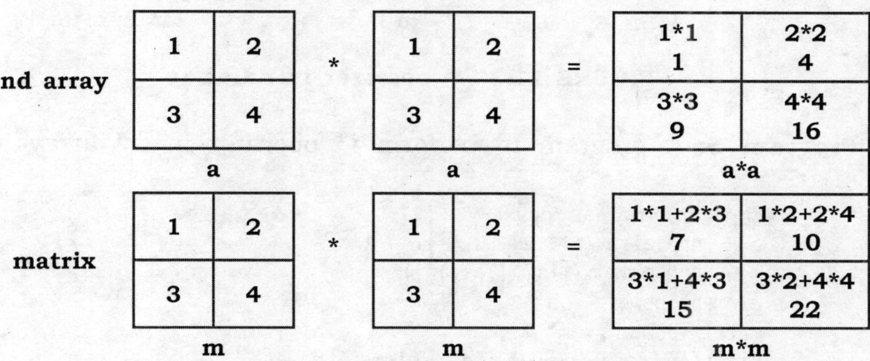

FIGURE 11.2   * operator in nd array

Program 4: A program to perform * operator in nd array and matrix.

```
import numpy as np9
a = np9.array([[1,2],[3,4]])
m = np9.matrix([[1,2],[3,4]])
element_mul = a*a
matrix_mul = m*m
print('ndarray multiplication :\n ',element_mul)
print('matrix multiplication :\n ',matrix_mul)
OUTPUT
C:\h_numpy>py h9.py
ndarray multiplication :
 [[ 1  4]
 [ 9 16]]
matrix multiplication :
 [[ 7 10]
 [15 22]]
```

** operator in nd array and matrix

- In case of nd array, ** operator performs power operation at element level.
- In case of matrix, ** operator performs power operation at matrix level.

$$m ** 2 ==> m * m$$

nd array	1	2	**	1	2	=	1**1 1	2**2 4
	3	4		3	4		3**3 9	4**4 16
	a			**a**			**a*a**	

matrix	1	2	**	1	2	=	1*1+2*3 7	1*2+2*4 10
	3	4		3	4		3*1+4*3 15	3*2+4*4 22
	m			**m**			**M**2=m*m**	

FIGURE 11.3   ** operator in nd array

Program 5: A program to perform ** operator in nd array and matrix.

```
import numpy as np9
a = np9.array([[1,2],[3,4]])
m = np9.matrix([[1,2],[3,4]])
element_power = a**2
matrix_power = m**2
print('ndarray power :\n ',element_power)
```

```
print('matrix power :\n ',matrix_power)
```
OUTPUT
```
C:\h_numpy>py h9.py
ndarray power :
 [[ 1  4]
  [ 9 16]]
matrix power :
 [[ 7 10]
  [15 22]]
```

T in nd array and matrix

In case of both nd array and matrix, T operator behaves in the same way.

Program 6: A program to perform T operator in nd array and matrix.

```
import numpy as np9
a = np9.array([[1,2],[3,4]])
m = np9.matrix([[1,2],[3,4]])
ndarray_T = a.T
matrix_T = m.T
print('ndarray transpose :\n ',ndarray_T)
print('matrix transpose :\n ',matrix_T)
```

OUTPUT
```
C:\h_numpy>py h9.py
ndarray transpose :
 [[1 3]
  [2 4]]
matrix transpose :
 [[1 3]
  [2 4]]
```

Conclusion

- The matrix class is the child class of nd array class. Hence all methods and properties of nd array class are by default available to the matrix class.
- We can use +, *, T, and ** operators for matrix objects also.
- In the case of nd array, the operator performs element level multiplication. But in the case of a matrix, the operator performs matrix multiplication.
- In the case of nd array, the operator performs power operation at the element level. But in the case of the class matrix, an operator performs 'matrix' power.
- Matrix class is always meant for 2-D array, only.
- It is no longer recommended to use.

Differences between nd array and matrix

nd array

- It can represent any n-dimension array.
- We can create it from any array_like object but not from a string.

- * operator is meant for element multiplication but not for the dot product.
- ** operator is meant for element level power operation.
- It is the parent class.
- It is recommended to use.

Matrix

- It can represent only a 2-dimension array.
- We can create it from either array_like object or from string.
- * operator is meant for dot product but not for the element multiplication.
- ** operator is meant for matrix power operation.
- It is the child class.
- It is not recommended to use.

11.3   MATRIX MULTIPLICATION BY USING dot() FUNCTION

To perform matrix level multiplication for a given two nd arrays, a,b, we have to use the dot() function. The dot() function is available in the NumPy module and nd array class.

For element level multiplication, we can perform a * b.

10	20
30	40

*

1	2
3	4

=

10*1 10	20*2 40
30*3 90	40*4 160

FIGURE 11.4   Matrix multiplication by using dot() function

Program 7: A program to print element level multiplication.

```
import numpy as np9
a = np9.array([[10,20],[30,40]])
b = np9.array([[1,2],[3,4]])
print("array a : \n ",a)
print("array b : \n ",b)
print("Element level multiplication ")
ele_multiplication = a* b
print("array b : \n ",ele_multiplication)
```

OUTPUT

```
C:\h_numpy>py h9.py
array a :
  [[10 20]
   [30 40]]
array b :
  [[1 2]
   [3 4]]
```

Element level multiplication

```
array b :
 [[  10  40]
  [  90 160]]
```

10	20		1	2		(10*1) + (20*3) 70	(10*2) + (20*4) 100
30	40	*	3	4	=	(30*1) + (40*3) 150	(30*2) + (40*4) 220

First row in a(0,20) * First column in b(1,3) (10*1) + (20*3)	First row in a(10,20) * Second column in b(2,4) (10*2) + (20*4)
Second row in a (30,40) * First column in b(1,3) (30*1) + (40*3)	Second row in a(30,40) * Second column in b(2,4) (30*2) + (40*4)

FIGURE 11.5   Matrix multiplication

Program 8: A program to print matrix multiplication using dot() method in nd array class.

```
import numpy as np9
a = np9.array([[10,20],[30,40]])
b = np9.array([[1,2],[3,4]])
dot_product = a.dot(b)
print("array a : \n ",a)
print("array b : \n ",b)
print("Matrix multiplication:\n ",dot_product)
```

OUTPUT

```
C:\h_numpy>py h9.py
array a :
 [[10 20]
  [30 40]]
array b :
 [[1 2]
  [3 4]]
Matrix multiplication:
 [[ 70 100]
  [150 220]]
```

EXERCISES

1. Write a program for creating matrix object from nested list.

2. Write a program to print matrix multiplication using dot() function in NumPy module.

12 LINEAR ALGEBRA FUNCTION FROM LINALG MODULE

12.1 INTRODUCTION

The numpy.linalg contains functions to perform linear algebra operations.

inv()	To find inverse of a matrix
matrix_power()	To find power of a matrix (like A^n)
det	To find the determinant of a matrix
solve()	To solve linear algebra equations

inv()

The inv() function is used to compute the (multiplicative) inverse of a matrix.

Inverse of 2-D array (matrix)

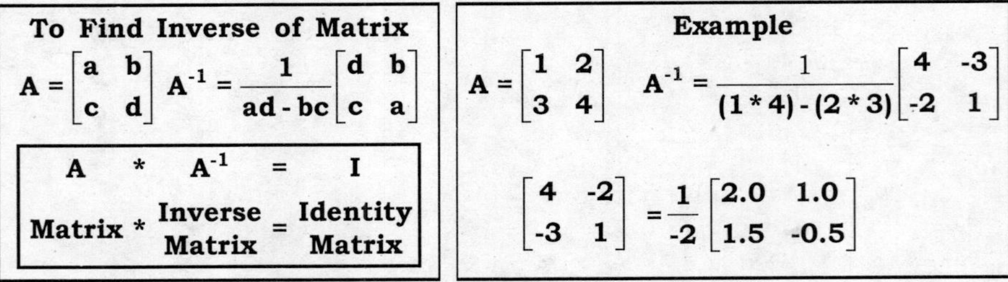

FIGURE 12.1   Inverse of 2-D array

Program 1: A program to find inverse of a matrix.

```
import numpy as np9
a = np9.array([[1,2],[3,4]])
ainv = np9.linalg.inv(a)
print("Original Matrix :: \n ",a)
print("Inverse of the Matrix :: \n ",ainv)
```

OUTPUT

```
C:\h_numpy>py h9.py
Original Matrix ::
 [[1  2]
  [3  4]]
Inverse of the Matrix ::
 [[-2.    1. ]
  [ 1.5 -0.5]]
```

To check the inverse of the matrix

```
dot(a,ainv) = dot(ainv,a) = eye(a,shape[0])
```

Program 2: A program to check the inverse of the matrix.

```
import numpy as np9
a = np9.array([[1,2],[3,4]])
ainv = np9.linalg.inv(a)
dot_produt = np9.dot(a,ainv)
i = np9.eye(2) # Identity matrix for the shape 2 ==> 2-D Arrray
print("Original Matrix :: \n ",a)
print("Inverse of the Matrix :: \n ",ainv)
print("Dot product of Matrix and Inverse of Matrix :: \n
",dot_produt)
print("Identity Matrix(2-D array using eye() function):: \n ",i)
```

OUTPUT

```
C:\h_numpy>py h9.py
Original Matrix ::
 [[1  2]
  [3  4]]
Inverse of the Matrix ::
 [[-2.    1. ]
  [ 1.5 -0.5]]
Dot product of Matrix and Inverse of Matrix ::
 [[1.0000000e+00  0.0000000e+00]
  [8.8817842e-16  1.0000000e+00]]
Identity Matrix(2-D array using eye() function)::
 [[1. 0.]
  [0. 1.]]
```

$$A = \begin{bmatrix} 1 & 2 \\ 3 & 4 \end{bmatrix} \qquad A^{-1} = \begin{bmatrix} -2.0 & 1.0 \\ 1.5 & -0.5 \end{bmatrix}$$

1*-2+2*1.5 -2+3	1*1+2*-0.5 1+-1
3*-2+4*1.5 -6+6	3*1+4*-0.5 3+2

$$A * A^{-1} \quad \begin{bmatrix} 1 & 2 \\ 3 & 4 \end{bmatrix} * \begin{bmatrix} -2.0 & 1.0 \\ 1.5 & -0.5 \end{bmatrix}$$

1	0
0	1

FIGURE 12.2   Dot matrix and inverse of matrix

12.2   np9.allclose()

This method is used to check whether the matrices are equal at element level.

Note:

- The results of floating point arithmetic varies from platform to platform.

Example

```
# example for allclose() function
import numpy as np9
a = np9.array([[1,2],[3,4]])
ainv = np9.linalg.inv(a)
dot_produt = np9.dot(a,ainv)
i = np9.eye(2)
print(np9.allclose(dot_produt,i))
```

OUTPUT

```
C:\h_numpy>py h9.py
True
```

CASE STUDY

We can find inverse only for square matrices, otherwise we will get error (LinAlgError).

```
# example for allclose() function
import numpy as np9
a = np9.arange(10).reshape(5,2)
print(np9.allclose(np9.linalg.inv(a)))
```

OUTPUT

```
C:\h_numpy>py h9.py
Traceback (most recent call last):
  File "C:\h_numpy\h9.py", line 5, in <module>
    print(np9.allclose(np9.linalg.inv(a)))
  File "<__array_function__ internals>", line 5, in inv
  File "C:\Users\Suresh S\AppData\Local\Programs\Python\Python39\
  lib\ site-packages\numpy\linalg\linalg.py", line 540, in inv
  _assert_stacked_square(a)
  File "C:\Users\Suresh S\AppData\Local\Programs\Python\Python39\
  lib\ site-packages\numpy\linalg\linalg.py", line 203, in _assert_
  stacked_square
  raise LinAlgError('Last 2 dimensions of the array must be
  square')
numpy.linalg.LinAlgError: Last 2 dimensions of the array must
be square
```

12.3   INVERSE OF 3-D ARRAY

3-D array is the collection of 2-D arrays. The computation of inverse of 3-D array involves the computation of inverse of every 2-D array.

Example

```
import numpy as np9
a = np9.arange(8).reshape(2,2,2)
ainv = np9.linalg.inv(a)
print("Original 3-D array :: \n ",a)
print("Inverse of 3-D array :: \n ",ainv)
```

OUTPUT

```
C:\h_numpy>py h9.py
Original 3-D array ::
 [[[0 1]
   [2 3]]
  [[4 5]
   [6 7]]]
Inverse of 3-D array ::
 [[[-1.5 0.5]
   [ 1.  0. ]]
  [[-3.5 2.5]
   [ 3. -2. ]]]
```

12.4 matrix_power()

To find power of a matrix,

$$matrix\_power(a,\ n)$$

if

n == 0 → Identity matrix

n > 0 → Normal power operation

n < 0 → First inverse and then power operation for absolute
value of n [abs(n)]

Program 3: A program where n=0 returns identity matrix.

```
import numpy as np9
a = np9.array([[1,2],[3,4]])
matrix_power = np9.linalg.matrix_power(a,0)
print("Original 2-D array(Matrix) :: \n ",a)
print("Matrix Power of 2-D array :: \n ",matrix_power)
```

OUTPUT

```
C:\h_numpy>py h9.py
Original 2-D array(Matrix) ::
 [[1 2]
  [3 4]]
Matrix Power of 2-D array ::
 [[1 0]
  [0 1]]
```

Program 4: A program where n < 0 returns first inverse operation
and then power operation.

```
import numpy as np9
a = np9.array([[1,2],[3,4]])
matrix_power = np9.linalg.matrix_power(a,-2)
```

```
print("Original 2-D array(Matrix) :: \n ",a)
print("Matrix Power of 2-D array :: \n ",matrix_power)
```

OUTPUT

```
C:\h_numpy>py h9.py
Original 2-D array(Matrix) ::
 [[1 2]
  [3 4]]
Matrix Power of 2-D array ::
 [[ 5.5 -2.5 ]
  [-3.75 1.75]]
```

Program 5: A program for the matrix power of inverse matrix and abs(n).

```
import numpy as np9
a = np9.array([[1,2],[3,4]])
ainv = np9.linalg.inv(a)
matrix_power = np9.linalg.matrix_power(ainv,2)
print("Original 2-D array(Matrix) :: \n ",a)
print("Inverse of Matrix :: \n ",ainv)
print("Matrix power of Inverse of Matrix and abs(n) :: \n
",matrix_power)
```

OUTPUT

```
C:\h_numpy>py h9.py
Original 2-D array(Matrix) ::
 [[1 2]
  [3 4]]
Inverse of Matrix ::
 [[-2. 1. ]
  [ 1.5 -0.5]]
Matrix power of Inverse of Matrix and abs(n) ::
 [[ 5.5 -2.5 ]
  [-3.75 1.75]]
```

a^{-2}	=	$(a^{-1})^2$ = $a^{-1} * a^{-1}$
a^{-2}	=	**np9.linalg.matrix_power(a,-2)**
$(a^{-1})^2$	=	**np9.linalg.matrix_power(aniv,2)** **aniv=np9.linalg.inv(s)**
$a^{-1} * a^{-1}$	=	**np9.linalg.matrix_power(aniv,2)** **aniv=np9.linalg.inv(s)**

FIGURE 12.3   Matrix power of inverse

CASE STUDY

We can find matrix_power only for square matrices, otherwise we will get error (LinAlgError).

```
import numpy as np9
a = np9.arange(10).reshape(5,2)
np9.linalg.matrix_power(a,2)
```

OUTPUT

```
C:\h_numpy>py h9.py
Traceback (most recent call last):
  File "C:\h_numpy\h9.py", line 3, in <module>
    np9.linalg.matrix_power(a,2)
  File "<__array_function__ internals>", line 5, in matrix_power
  File "C:\Users\Suresh S\AppData\Local\Programs\Python\Python39\
  lib\ site-packages\numpy\linalg\linalg.py", line 620, in matrix_
  power
  _assert_stacked_square(a)
  File "C:\Users\Suresh S\AppData\Local\Programs\Python\Python39\
  lib\ site-packages\numpy\linalg\linalg.py", line 203, in _assert_
  stacked_square
    raise LinAlgError('Last 2 dimensions of the array must be
  square')
numpy.linalg.LinAlgError: Last 2 dimensions of the array must
be square
```

12.5 det()

It is used to compute the determinant of an array.

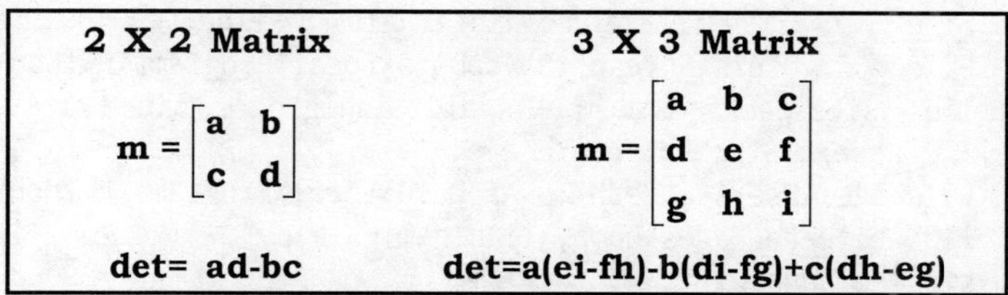

FIGURE 12.4 Determinant of matrix

Example

```
import numpy as np9
a = np9.array([[1,2],[3,4]])
adet = np9.linalg.det(a)
print("Original Matrix : \n ",a)
print("Determinant of Matrix : \n ",adet)
```

OUTPUT

```
C:\h_numpy>py h9.py
Original Matrix :
 [[1 2]
 [3 4]]
Determinant of Matrix :
 -2.0000000000000004
```

Example

```
import numpy as np9
a = np9.arange(9).reshape(3,3)
adet = np9.linalg.det(a)
```

```
print("Original Matrix : \n ",a)
print("Determinant of Matrix : \n ",adet)
```

OUTPUT

```
C:\h_numpy>py h9.py
Original Matrix :
 [[0 1 2]
  [3 4 5]
  [6 7 8]]
Determinant of Matrix :
 0.0
```

CASE STUDY

We can find determinant only for square matrices, otherwise we will get error (LinAlgError).

```
import numpy as np9
a = np9.arange(10).reshape(5,2)
print(np9.linalg.det(a))
OUTPUT
C:\h_numpy>py h9.py
Traceback (most recent call last):
 File "C:\h_numpy\h9.py", line 3, in <module>
  print(np9.linalg.det(a))
 File "<__array_function__ internals>", line 5, in det
 File "C:\Users\Suresh S\AppData\Local\Programs\Python\Python39\
 lib \site-packages\numpy\linalg\linalg.py", line 2155, in
det _assert_stacked_square(a)
 File "C:\Users\Suresh S\AppData\Local\Programs\Python\Python39\
 lib\ site-packages\numpy\linalg\linalg.py", line 203, in _assert_
 stacked_square
 raise LinAlgError('Last 2 dimensions of the array must be
 square')
numpy.linalg.LinAlgError: Last 2 dimensions of the array must
be square
```

12.6   solve()

It is used to solve linear algebra equations.

$$solve(a, b)$$

where

   a : (..., M, M) array_like ==> Coefficient matrix.
   b : {(..., M,), (..., M, K)}, array_like ==> Ordinate or "dependent variable" values.

2-variables

CASE STUDY

Problem

   • Boys and girls attended Durga sir's data science class.

- For boys, fee is $3 and for girls, fee is $8.
- For a certain batch, 2200 people attended, and $10100 fee was collected.
- How many boys and girls attended for that batch?

<table>
<tr><td>

Assume that

x= number of boys

y= number of girls

Total Students =2200(x+y)

Total Fee Collected =10100(3x+8y)

x+y =2200 -->1

3x+8y =10100 -->2

x =2200-y -->3

substitute x value in eq -2 it will becomes

3(2200-y) + 8y= 10100 -->4

6600-3y+8y =10100

5y=10100-6600

y=3500/5 =700

Substitute the value in eq-1 we will get x value

x+700=2200

x=2200-700=1500

No of Boys (x) : 1500

No of girls (y) : 700

</td><td>

x+y=2200

3x+8y=10100

Coefficient Matrix		**Coordinates**
x	**y**	
1	**1**	**[1500, 10100]**
3	**8**	

coef=np9.array([[1,1],[3,8]])

ord=np9.array([2200.10100])

np9.linalg.solve(coef.ord)

array([1500,700])

</td></tr>
</table>

FIGURE 12.5   Solution of 2-variable problem

```
# solution
# x+y = 2200
# 3x+8y=10100
import numpy as np9
coef = np9.array([[1,1],[3,8]])
dep = np9.array([2200,10100])
result = np9.linalg.solve(coef,dep)
print("Coefficient Matrix : \n ",coef)
print("Dependent Matrix : \n ",dep)
print("Solution array : ",result)
print("Type of result : ",type(result))
```

OUTPUT

```
C:\h_numpy>py h9.py
Coefficient Matrix :
 [[1 1]
 [3 8]]
Dependent Matrix :
 [ 2200 10100]
 Solution array : [1500. 700.]
Type of result : <class 'numpy.ndarray'>
```

3 variables

<table>
<tr><td>

-4x +7y -2z = 2

x - 2y + z = 3

2x -3y +z = -4

</td><td>

Coefficient Matrix

x	y	z
[-4	7	-2
1	-2	1
2	-3	1]

</td><td>

Ordinate/Dependent Variable

[2,   3,   -4]

</td></tr>
</table>

```
coef=np9.array([[-4,7,-2],[1,-2,1],[2,-3,1]])
dep=np9.array([2,3,-4])
Result -np9.linalg.solve(coef.dep)
```

FIGURE 12.6   Solution of 3-variable Problem

Program 6: A program for finding the values of the 3 variables.

```
# -4x+7y-2z = 2
# x-2y+z = 3
# 2x-3y+z = -4
import numpy as np9
coef = np9.array([[-4,7,-2],[1,-2,1],[2,-3,1]])
dep = np9.array([2,3,-4])
result = np9.linalg.solve(coef,dep)
print("Coefficient Matrix : \n ",coef)
print("Dependent Matrix : \n ",dep)
print("Solution array : ",result)
```

OUTPUT

```
C:\h_numpy>py h9.py
Coefficient Matrix :
 [[-4 7 -2]
  [ 1 -2 1]
  [ 2 -3 1]]
Dependent Matrix :
 [ 2 3 -4]
Solution array :  [-13. -6. 4.]
```

EXERCISES

1. Write a program where n > 0 and it returns normal power operation.

2. Write a program where the dot product of inverse matrix, i.e., dot(ainv) * dot(ainv) = result of n < 0.

13

UNIQUE FUNCTION AND IO OPERATIONS WITH NUMPY

13.1 unique() FUNCTION

The unique() function is used to find the unique items and their count in nd array. It returns the sorted unique elements of an array.

Syntax

```
unique(ar, return_index=False, return_inverse=False, return_counts=False,
                                                              axis=None)
```

There are three optional outputs in addition to the unique elements:
* The indices of the input array that give the unique values ==> return_index.
* The indices of the unique array that reconstruct the input array.
* The number of times each unique value comes up in the input array ==> return_counts.

Program 1: A program to get array with unique elements.

```
import numpy as np9
a = np9.array([1,1,2,3,4,2,3,4,4,1,2,3,4,5,5,6])
print("Original array : ",a)
print("Unique elements in the array using np9.unique(a) : ",np9.
unique(a))
```

OUTPUT

```
C:\h_numpy>py h9.py
Original array : [1 1 2 3 4 2 3 4 4 1 2 3 4 5 5 6]
Unique elements in the array using np9.unique(a) : [1 2 3 4 5 6]
```

Program 2: A program to get indices of unique elements.

```
import numpy as np9
a = np9.array([1,1,2,3,4,2,3,4,4,1,2,3,4,5,5,6])
```

```
items,indices = np9.unique(a,return_index=True)
print("Original array : ",a)
print("Unique Elements :",items)
print("indices of unique elements: ",indices)
```

OUTPUT

```
C:\h_numpy>py h9.py
Original array : [1 1 2 3 4 2 3 4 4 1 2 3 4 5 5 6]
Unique Elements : [1 2 3 4 5 6]
indices of unique elements: [ 0 2 3 4 13 15]
```

Program 3: A program to get count of unique elements.

```
import numpy as np9
a = np9.array([1,1,2,3,4,2,3,4,4,1,2,3,4,5,5,6])
items,counts = np9.unique(a,return_counts=True)
print("Original array : ",a)
print("Unique Elements :",items)
print("count of unique elements: ",counts)
```

OUTPUT

```
C:\h_numpy>py h9.py
Original array : [1 1 2 3 4 2 3 4 4 1 2 3 4 5 5 6]
Unique Elements : [1 2 3 4 5 6]
count of unique elements: [3 3 3 4 2 1]
```

Program 4: A program to get the count of all unique elements in the given array.

```
import numpy as np9
a = np9.array([1,1,2,3,4,2,3,4,4,1,2,3,4,5,5,6])
items,indices,counts = np9.unique(a,return_index=True,return_
counts=True)
print("Original array : ",a)
print("Unique Elements :",items)
print("indices of unique elements: ",indices)
print("count of unique elements: ",counts)
```

OUTPUT

```
C:\h_numpy>py h9.py
Original array : [1 1 2 3 4 2 3 4 4 1 2 3 4 5 5 6]
Unique Elements : [1 2 3 4 5 6]
indices of unique elements: [ 0 2 3 4 13 15]
count of unique elements: [3 3 3 4 2 1]
```

13.2 IO OPERATIONS

We can save/write nd array objects to a binary file for future purpose. Whenever these objects are required, we can read from that binary file.

- **save()**   To save/write nd array object to a file
- **load()**   To read nd array object from a file

Syntax

```
save(file, arr, allow_pickle=True, fix_imports=True)
```

Save an array to a binary file in NumPy .np9y format.

Load arrays or pickled objects from .np9y, .np9z or pickled files.

```
load(file, mmap_mode=None, allow_pickle=False, fix_imports=True,
                                              encoding='ASCII')
```

Program 5: A program for saving single nd array object to a file and read back: (save_read_obj.py)

```
import numpy as np9
a = np9.array([[10,20,30],[40,50,60]]) #2-D array with shape:(2,3)
#save/serialize ndarray object to a file
np9.save('out.np9y',a)
#load/deserialize ndarray object from a file
out_array = np9.load('out.np9y')
print(out_array)
```

OUTPUT

```
C:\h_numpy>py h9.py
 [[10 20 30]
  [40 50 60]]
```

Note:

- The data will be stored in the binary form.
- File extension should be .np9y, otherwise save() function itself will add that extension.
- By using save() function, we can write only one object to the file. If we want to write multiple objects to a file then we should go for savez() function.

PRACTICE PROGRAMS

Program 6: A program for saving mulitple nd array objects to the binary file:

```
import numpy as np9
a = np9.array([[10,20,30],[40,50,60]]) #2-D array with shape:(2,3)
b = np9.array([[70,80],[90,100]]) #2-D array with shape:(2,2)
#save/serialize ndarrays object to a file
np9.savez('out.np9z',a,b)
#reading ndarray objects from a file
np9zfileobj = np9.load('out.np9z') #returns Np9zFile object
print("Type of the np9zfileobj : ",type(np9zfileobj))
print(np9zfileobj.files)
print(np9zfileobj['arr_0'])
print(np9zfileobj['arr_1'])
```

OUTPUT

```
C:\h_numpy>py h9.py
```

```
Type of the np9zfileobj : <class 'numpy.lib.np9yio.Np9zFile'>
 ['arr_0', 'arr_1']
[[10 20 30]
 [40 50 60]]

[[ 70 80]
 [ 90 100]]
```

Program 7: A program for reading the file objects using for loop.

```python
import numpy as np9
a = np9.array([[10,20,30],[40,50,60]]) #2-D array with shape:(2,3)
b = np9.array([[70,80],[90,100]]) #2-D array with shape:(2,2)
#save/serialize ndarrays object to a file
np9.savez('out.np9z',a,b)
#reading ndarray objects from a file
np9zfileobj = np9.load('out.np9z') #returns Np9zFile object ==>
<class 'numpy.lib.np9yio.Np9zFile'>
print(type(np9zfileobj))
print(np9zfileobj.files)
for i in np9zfileobj:
    print("Name of the file : ",i)
    print("Contents in the file :")
    print(np9zfileobj[i])
    print("*"*80)
```

OUTPUT

```
C:\h_numpy>py h9.py
<class 'numpy.lib.np9yio.Np9zFile'>
['arr_0', 'arr_1']
Name of the file : arr_0
Contents in the file :
[[10 20 30]
 [40 50 60]]
********************************************************************************
Name of the file : arr_1
Contents in the file :
[[ 70 80]
 [ 90 100]]
```

Note:

- np9.save() is used to save an array to a binary file in .np9y format.
- np9.savez() is used to save several arrays into a single file in .np9z format but in the uncompressed form.
- np9.savez_compressed() is used to save several arrays into a single file in .np9z format but in the compressed form.
- np9.load() is used to load/read arrays from .np9y or .np9z files.

savez_compressed() and load() => nd array compressed form

Program 8: A program for compressing the file.

```python
import numpy as np9
a = np9.array([[10,20,30],[40,50,60]]) #2-D array with shape:(2,3)
b = np9.array([[70,80],[90,100]]) #2-D array with shape:(2,2)
#save/serialize ndarrays object to a file
np9.savez_compressed('out_compressed.np9z',a,b)
#reading ndarray objects from a file
np9zfileobj = np9.load('out_compressed.np9z') #returns Np9zFile
object
#print(type(np9zfileobj))
print(np9zfileobj.files)
print(np9zfileobj['arr_0'])
print(np9zfileobj['arr_1'])
```

OUTPUT

```
C:\h_numpy>py h9.py
  ['arr_0', 'arr_1']
 [[10  20  30]
  [40  50  60]]
 [[ 70  80]
  [ 90 100]]
```

Now, a question arises that when we can save object in the compressed form, then what is the need of the uncompressed form?

Well, in the compressed form, the memory will be saved, but performance is not up to the mark, whereas in uncompressed form, the memory won't be saved, but the performance wise, it is good.

Note:

- If we are using save() function, the file extension is np9y.
- If we are using savez() or savez_compressed() functions, the file extension is np9z.

savetxt() and loadtxt() => ndarray object to text file

- To save nd array object to a text file, we will use **savetxt()** function.
- To read nd array object from a text file, we will use **loadtxt()** function.

Example

```python
import numpy as np9
a = np9.array([[10,20,30],[40,50,60]]) #2-D array with shape:(2,3)
#save/serialize ndarrays object to a file
np9.savetxt('out.txt',a,fmt='%.1f')
#reading ndarray objects from a file and default dtype is float
out_array1 = np9.loadtxt('out.txt')
print("Output array in default format : float")
print(out_array1)
#reading ndarray objects from a file and default dtype is int
print("Output array in int format")
out_array2 = np9.loadtxt('out.txt',dtype=int)
print(out_array2)
```

OUTPUT

```
C:\h_numpy>py h9.py
Output array in default format : float
[[10. 20. 30.]
 [40. 50. 60.]]
Output array in int format
[[10 20 30]
 [40 50 60]]
```

savetxt() conclusion

By using savetxt(), we can store nd array of type 1-D and 2-D only. If we use 3-D array to store in a file, then it will give an error.

Program 9: A program to store str nd array in a file, we must specify the fmt parameter as str.

```
import numpy as np9
a1 = np9 array([['Sunny',1000],['Bunny',2000],['Chinny',3000],
['Pinny',4000]])
#save this ndarray to a text file
np9.savetxt('out.txt',a1,fmt='%s %s')
#reading ndarray from the text file
a2 = np9.loadtxt('out.txt',dtype='str')
print("Type of a2(fetching the data from text file) : ",type(a2))
print("The a2 value after fetching the data from text file :
\n ",a2)
a2 = np9.loadtxt('out.txt',dtype='str')
print("Type of a2 : ",type(a2))
print(a2)
```

OUTPUT

```
C:\h_numpy>py h9.py
Type of a2(fetching the data from text file) : <class 'numpy.
ndarray'>
The a2 value after fetching the data from text file :
 [['Sunny' '1000']
 ['Bunny' '2000']
 ['Chinny' '3000']
 ['Pinny' '4000']]
Type of a2 : <class 'numpy.ndarray'>
 [['Sunny' '1000']
 ['Bunny' '2000']
 ['Chinny' '3000']
 ['Pinny' '4000']]
```

Summary

- Save one nd array object to the binary file (save() and load()).
- Save multiple nd array objects to the binary file in uncompressed form (savez() and load()).
- Save multiple nd array objects to the binary file in compressed form (savez_compressed() and load()).

- Save nd arry object to the text file (savetxt() and loadtxt()).
- Save nd arry object to the csv file (savetxt() and loadtxt() with delimiter=',').

EXERCISES

1. Write a program to get the count of all letters in the given array.

2. Write a program to display nd array objects to the csv file (csv stands for comma separated values).

14 BASIC STATISTICS WITH NUMPY

14.1 INTRODUCTION

In the data science domain, we may require to collect, store and analyze huge amount of data. From this data, we may require to find some basic statistics such as:

- Minimum value
- Maximum value
- Average value
- Sum of all values
- Mean value
- Median value
- Variance
- Standard deviation etc.

14.2 MINIMUM VALUE

- np9.min(a)
- np9.amin(a)
- a.min()

1-D array

Program 1: A program to print minimum value in 1-D array.

```python
import numpy as np9
a = np9.array([10,5,20,3,25])
print("1-D array : ",a)
print("np9.min(a) value : ",np9.min(a))
print("np9.amin(a) value : ",np9.amin(a))
print("a.min() value : ",a.min())
```

OUTPUT

```
C:\h_numpy>py h9.py
1-D array : [10 5 20 3 25]
np9.min(a) value : 3
np9.amin(a) value : 3
a.min() value : 3
```

2-D Array

- **axis=None (default)** – The array is flattened to 1-D array and then find the minimum value.
- **axis=0** – Minimum row with 3 elements
- **axis=1** – Minimum column with 4 elements
- **axis=0** – Minimum row considering all the columns in that min row value
- **axis=1** – Minimum column considering all rows in that min column value

min value along axis=0 and axis=1

100	20	30	100	20	30
10	50	60	10	50	60
25	15	18	25	15	18
4	5	19	4	5	19

axis=0	axis=1
[4,5,18]	[20,10,15,4]

FIGURE 14.1   Minimum value along axis=0 and axis=1

Program 2: A program to print minimum value in 2D array.

```
import numpy as np9
a = np9.array([[100,20,30],[10,50,60],[25,15,18],[4,5,19]])
print("array a : \n ",a)
print("Minimum value along axis=None : ",np9.min(a))
print("Minimum value along axis-0 : ",np9.min(a,axis=0))
print("Minimum value along axis-1 : ",np9.min(a,axis=1))
```

OUTPUT

```
C:\h_numpy>py h9.py
array a :
 [[100 20 30]
 [ 10 50 60]
 [ 25 15 18]
 [ 4  5 19]]
Minimum value along axis=None : 4
Minimum value along axis-0 : [ 4 5 18]
Minimum value along axis-1 : [20 10 15 4]
```

Program 3: A program to print minimum value along axes from an array.

```
import numpy as np9v
a = np9.arange(24)
np9.random.shuffle(a)
a = a.reshape(6,4)
print("array a : \n ",a)
print("Minimum value along axis=None : ",np9.min(a))
print("Minimum value along axis-0 : ",np9.min(a,axis=0))
print("Minimum value along axis-1 : ",np9.min(a,axis=1))
```

OUTPUT

```
C:\h_numpy>py h9.py
array a :
 [[ 3  6  1 13]
  [ 8 23  0 17]
  [ 4 15 18 16]
  [ 2 20 19 10]
  [ 7 12 14 21]
  [22  5 11  9]]
Minimum value along axis=None : 0
Minimum value along axis-0 : [2 5 0 9]
Minimum value along axis-1 : [1 0 4 2 7 5]
```

14.3   MAXIMUM VALUE

- np9.max(a)
- np9.amax(a)
- a.max()

1-D array

Program 4: A program to display maximum value from 1-D array.

```
import numpy as np9
a = np9.array([10,5,20,3,25])
print("1-D array : ",a)
print("np9.max(a) value : ",np9.max(a))
print("np9.amax(a) value : ",np9.amax(a))
print("a.max() value : ",a.max())
```

OUTPUT

```
C:\h_numpy>py h9.py
1-D array : [10 5 20 3 25]
np9.max(a) value : 25
np9.amax(a) value : 25
a.max() value : 25
```

2-D array

- **axis=None (default)** – The array is flattened to 1-D array and then find the maximum value.

- **axis=0** – Maximum row with 3 elements.
- **axis=1** – Maximum column with 4 elements.
- **axis=0** – Maximum row considering all the columns in that max row value.
- **axis=1** – Maximum column considering all rows in that max column value.

max value along axis=0 and axis=1

100	20	30
10	50	60
25	15	18
4	5	19

100	20	30
10	50	60
25	15	18
4	5	19

axis=0   axis=1

[100,50,60]   [100,60,25,19]

FIGURE 14.2   Maximum value along axis=0 and axis=1

Program 5: A program to print maximum value from given 2-D array.

```
import numpy as np9
a = np9.array([[100,20,30],[10,50,60],[25,15,18],[4,5,19]])
print("array a : \n ",a)
print("Maximum value along axis=None : ",np9.max(a))
print("Maximum value along axis-0 : ",np9.max(a,axis=0))
print("Maximum value along axis-1 : ",np9.max(a,axis=1))
```

OUTPUT

```
C:\h_numpy>py h9.py
array a :
 [[100 20 30]
 [ 10 50 60]
 [ 25 15 18]
 [  4  5 19]]
Maximum value along axis=None : 100
Maximum value along axis-0 : [100 50 60]
Maximum value along axis-1 : [100 60 25 19]
```

14.4   SUM OF THE ELEMENTS

- np9.sum()
- a.sum()

Program 6: A program to print sum of elements of 1-D array.

```
import numpy as np9
a = np9.arange(4)
print("The array a : ",a)
```

```
print("sum of elements using np9.sum(a) :: ",np9.sum(a))
print("sum of elements using a.sum() :: ",a.sum())
```

OUTPUT

```
C:\h_numpy>py h9.py
The array a : [0 1 2 3]
sum of elements using np9.sum(a) :: 6
sum of elements using a.sum() :: 6
```

Sum of elements along axis=0 and axis=1

<table>
<tr><td>0</td><td>1</td><td>2</td></tr>
<tr><td>3</td><td>4</td><td>5</td></tr>
<tr><td>6</td><td>7</td><td>8</td></tr>
</table>

axis=0

[9,12,15]

<table>
<tr><td>0</td><td>1</td><td>2</td></tr>
<tr><td>3</td><td>4</td><td>5</td></tr>
<tr><td>6</td><td>7</td><td>8</td></tr>
</table>

axis=1

[3,12,21]

FIGUER 14.3   Sum of elements along axis=0 and axis=1

Program 7: A program to print sum of elements in an array.

```
import numpy as np9
a = np9.arange(9).reshape(3,3)
print("array a : \n ",a)
print("Sum along axis=None : ",np9.sum(a))
print("Sum along axis-0 : ",np9.sum(a,axis=0))
print("Sum along axis-1 : ",np9.sum(a,axis=1))
```

OUTPUT

```
C:\h_numpy>py h9.py
array a :
 [[0 1 2]
  [3 4 5]
  [6 7 8]]
Sum along axis=None : 36
Sum along axis-0 : [ 9 12 15]
Sum along axis-1 : [ 3 12 21]
```

14.5   MEAN VALUE

Mean is the sum of elements along the specified axis divided by the number of elements.

- np9.mean(a)
- a.mean()

Program 8: A program to print mean value for given 1-D array.

```
import numpy as np9
a = np9.arange(5)
print("1-D array : ",a)
```

```
print("np9.mean(a) value : ",np9.mean(a))
print("a.mean() value : ",a.mean())
```
OUTPUT
```
C:\h_numpy>py h9.py
1-D array : [0 1 2 3 4]
np9.mean(a) value : 2.0
a.mean() value : 2.0
```

2-D array

- **axis=None (default)** – The array is flattened to 1-D array and then find the mean (average) value.
- **axis=0** – rows. Consider columns with all rows and find the average.
- **axis=1** – columns. Consider rows with all columns and find the average.

mean value along axis=0 and axis=1

axis=0

0	1	2
3	4	5
6	7	8

(0+3+6)/3	(1+4=7)/3	(2+5+8)/3
3.0	4.0	5.0

axis=1

0	1	2		(0+1+2)/3	1.0
3	4	5		(3+4+5)/3	4.0
6	7	8		(6+7+8)/3	7.0

FIGURE 14.4   Mean value along axis=0 and axis=1

14.6   MEDIAN VALUE

Median means the middle element of the array (sorted form). If the array contains an even number of elements, then the median is the middle element value. If the array contains an odd number of elements, then the median is the average of the two middle element values.

```
np9.median(a)
```

Program 9: A program to print median for the given array with even and odd number elements.

```
import numpy as np9
a = np9.array([10,20,30,40])
b = np9.array([10,20,30,40,50])
print("The array with even number of elements :",a)
print("Median of the array with even number of elements : ",np9.
median(a))
print()
```

```
print("The array with odd number of elements : ",b)
print("Median of the array with odd number of elements : ",np9.
median(b))
```

OUTPUT

```
C:\h_numpy>py h9.py
The array with even number of elements : [10 20 30 40]
Median of the array with even number of elements : 25.0
The array with odd number of elements : [10 20 30 40 50]
Median of the array with odd number of elements : 30.0
```

Program 10: A program to print unsorted array (even no of elements) that is converted to sorted array and then median is calculated.

```
import numpy as np9
a = np9.array([80,20,60,40])
print("The array with even number of elements(unsorted) : ",a)
print("*"*100)
print("This step is calculated internally ")
print("sorted form of given array : ",np9.sort(a))
print("*"*100)
print("Median of the array with even number of elements : ",np9.
median(a))
```

OUTPUT

```
C:\h_numpy>py h9.py
The array with even number of elements(unsorted) : [80 20 60 40]
********************************************************************
This step is calculated internally
sorted form of given array : [20 40 60 80]
********************************************************************
Median of the array with even number of elements : 50.0
```

2-D array

- **axis=None (default)** – The array is flattened to 1-D array (sorted) and find the median value.
- **axis=0** – rows. Consider columns with all rows and find the median.
- **axis=1** – columns. Consider rows with all columns and find the median.

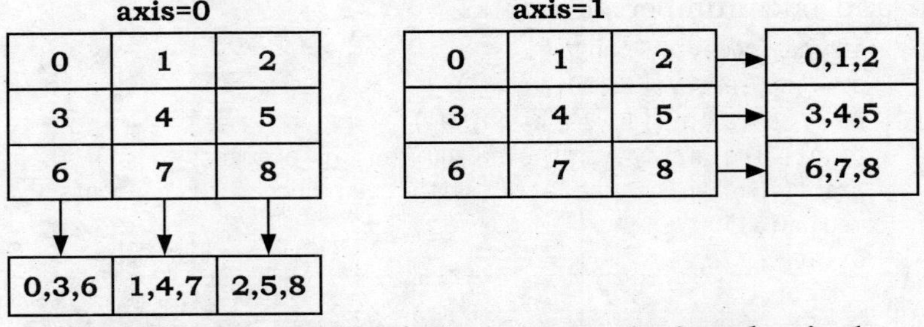

FIGURE 14.5 Median value along axis=0 and axis=1

Program 11: A program to print median of the given 2-D array elements.

```
import numpy as np9
a = np9.arange(9).reshape(3,3)
print("The original 2-D array(already sorted) : \n ",a)
print("Mean of the 2-D array along axis=None : ",np9.median(a))
print("Mean of the 2-D array along axis=0 : ",np9.median(a,axis=0))
print("Mean of the 2-D array along axis=1 : ",np9.median(a,axis=1))
```

OUTPUT

```
C:\h_numpy>py h9.py
The original 2-D array(already sorted) :
 [[0 1 2]
  [3 4 5]
  [6 7 8]]
Mean of the 2-D array along axis=None : 4.0
Mean of the 2-D array along axis=0 : [3. 4. 5.]
Mean of the 2-D array along axis=1 : [1. 4. 7.]
```

median along axis=0 and axis=1

axis=0		
22	55	88
11	44	77
33	66	99

axis=1			sorted	
22	55	88	22,55,88	22,55,88
11	44	77	11,44,77	11,44,77
33	66	99	33,66,99	33,66,99

22,11,33	55,44,66	88,77,99

sorted

11,22,33	44,55,66	77,88,99

FIGURE 14.6   Median along axis=0 and axis=1

Program 12: A program to print the median of the given 2-D array with unsorted elements.

```
import numpy as np9
a = np9.array([[22,55,88],[11,44,55],[33,66,99]])
print("The original 2-D array(unsorted) : \n ",a)
print("Mean of the 2-D array along axis=None : ",np9.median(a))
print("Mean of the 2-D array along axis=0 : ",np9.median(a,axis=0))
print("Mean of the 2-D array along axis=1 : ",np9.median(a,axis=1))
```

OUTPUT

```
C:\h_numpy>py h9.py
The original 2-D array(unsorted) :
 [[22 55 88]
  [11 44 55]
  [33 66 99]]
```

```
Mean of the 2-D array along axis=None : 55.0
Mean of the 2-D array along axis=0 : [22. 55. 88.]
Mean of the 2-D array along axis=1 : [55. 44. 66.]
```

axis=0		**axis=0**		
array([[0, 7, 1],	array([[0, 7, 1],	0,7,1	0,1,7
[4, 3, 8],	[4, 3, 8],	4,3,8	3,4,8
[5, 6, 2],])	[5, 6, 2],])	5,6,2	2,5,6
0,4,5 7,3,6 1,8,2			
0,4,5 3,7,6 1,2,8			

FIGURE 14.7   Median of unsorted elements

14.7   VARIANCE VALUE

The variance is a measure of variability. It is calculated by taking the average of squared deviations from the mean.

- `np9.var(a)`
- `a.var()`

Original Array (x)	Deviation (x-Mean)	Square of Deviation
1	1-3=-2	4
2	2-3=-1	1
3	3-3=0	0
4	4-3=1	1
5	5-3=2	4

> **mean(a) =(1+2+3+5+6)/5 =3.0**
> **deviation from the mean:**
> **[-2.0,-1.0,0.0,1.0,2.0] squares**
> **of deviations from the mean :**
> **[4.0,1.0,0.0,1.0,4.0]**
> **Average of square of deviation**
> **from the mean : (4+1+0+1+4)/5=2.0**
> **VARIANCE**

FIGURE 14.8   Variance of an array

Program 13: A program to print variance of the given 1-D array.

```python
import numpy as np9
a = np9.array([1,2,3,4,5])
print("Original 1-D array : ",a)
print("Variance of 1-D array unsing np9.var(a): ",np9.var(a))
print("Variance of 1-D array unsing a.var(): ",a.var())
```

OUTPUT

```
C:\h_numpy>py h9.py
Original 1-D array : [1 2 3 4 5]
Variance of 1-D array unsing np9.var(a): 2.0
Variance of 1-D array unsing a.var(): 2.0
```

2-D array

- **axis=None (default)** – The array is flattened to 1-D array (sorted) and the variance value is calculated.
- **axis=0** – rows. Consider columns with all rows and find the variance.
- **axis=1** – columns. Consider rows with all columns and find the variance.

Program 14: A program to print variance of the given 2-D array.

```
import numpy as np9
a = np9.arange(6).reshape(2,3)
print("Original 2-D array :\n {a}")
print("Variance of 2-D array using np9.var(a) along axis=None:
",np9.var(a))
print("Variance of 2-D array using np9.var(a) along axis=0:
",np9.var(a,axis=0))
print("Variance of 2-D array using np9.var(a) along axis=1:
",np9.var(a,axis=1))
```

OUTPUT

```
C:\h_numpy>py h9.py
Original 2-D array :
{a}
Variance of 2-D array using np9.var(a) along axis=None:
2.9166666666666665
Variance of 2-D array using np9.var(a) along axis=0: [2.25 2.25
2.25]
Variance of 2-D array using np9.var(a) along axis=1: [0.66666667
0.66666667]
```

14.8   STANDARD DEVIATION VALUE

Standard deviation is the square root of the variance (average of squares of deviations from the mean).

- np9.std(a)
- a.std()

Program 15: A program to print standard deviation for the given 1-D array.

```
import math
import numpy as np9
a = np9.array([1,2,3,4,5])
print("Original 1-D array : ",a)
print("Variance of 1-D array unsing np9.var(a): ",np9.var(a))
print("Standard Deviation of 1-D array unsing np9.std(a): ",np9.
std(a))
print("Square root of Variannce : ",math.sqrt(np9.var(a)))
```

OUTPUT

```
C:\h_numpy>py h9.py
Original 1-D array : [1 2 3 4 5]
Variance of 1-D array unsing np9.var(a): 2.0
Standard Deviation of 1-D array unsing np9.std(a): 1.4142135623730951
Square root of Variannce :  1.4142135623730951
```

Program 16: A program to print standard deviation for the given 1-D array.

```
import math
import numpy as np9
a = np9.arange(6).reshape(2,3)
print("Original 2-D array :\n ",a)
print("*"*100)
print("Variance of 2-D array using np9.var(a) along axis=None: ",np9.var(a))
print("Std Deviation of 2-D array using np9.std(a) along axis=None: ",np9.std(a))
print("Square root of Variannce : ",math.sqrt(np9.var(a)))
print("*"*100)
print("Variance of 2-D array using np9.var(a) along axis=0: ",np9.var(a,axis=0))
print("Std Deviation of 2-D array using np9.std(a) along axis=0:",np9.std(a,axis=0))
print("*"*100)
print("Variance of 2-D array using np9.var(a) along axis=1: ",np9.var(a,axis=1))
print("Std Deviation of 2-D array using np9.std(a) along axis=1:",np9.std(a,axis=1))
print("*"*100)
```

OUTPUT

```
C:\h_numpy>py h9.py
Original 2-D array :
  [[0 1 2]
   [3 4 5]]
*******************************************************************
Variance of 2-D array using np9.var(a) along axis=None:
2.9166666666666665
Std Deviation of 2-D array using np9.std(a) along axis=None:
1.707825127659933
Square root of Variannce : 1.707825127659933
*******************************************************************
Variance of 2-D array using np9.var(a) along axis=0: [2.25 2.25
2.25]
Std Deviation of 2-D array using np9.std(a) along axis=0: [1.5
1.5 1.5]
*******************************************************************
Variance of 2-D array using np9.var(a) along axis=1: [0.66666667
0.66666667]
```

```
Std Deviation  of  2-D  array  using  np9.std(a)  along  axis=1:
[0.81649658     0.81649658]
```

```
******************************************************************
```

Summary

- **np9.min(a)/np9.amin(a)/a.min()**→Returns the minimum value of the array.
- **np9.max(a)/np9.amax(a)/a.max()**→Returns the maximum value of the array.
- **np9.sum(a)/a.sum()**→Returns the sum of values of the array.
- **np9.mean(a)/a.mean()**→Returns the arithmetic mean of the array.
- **np9.median(a)**→Returns median value of the array.
- **np9.var(a)/a.var()**→Returns variance of the values in the array.
- **np9.std(a)/a.std()**→Returns standard deviation of the values in the array.

EXERCISES

1. Write a program to print minimum values along axes for the 3-D array.

2. Write a program to print mean value along axes for the given 2-D array.

3. Write a program to print unsorted array (odd no. of elements) that is converted to sorted array and then median is calculated.

4. Write a program to display median of the unsorted elements in 2-D array using shuffle.

15 NUMPY MATHEMATICAL AND FINANCIAL FUNCTIONS

15.1 MATHEMATICAL FUNCTIONS

The functions which operate element by element on the whole array are called universal functions. To perform mathematical operations, NumPy library contains several universal functions (ufunc).

np9.exp(a)	Takes e to the power of each value (e value: 2.7182)
np9.sqrt(a)	Returns square root of each value.
np9.log(a)	Returns logarithm of each value.
np9.sin(a)	Returns the sine of each value.
np9.cos(a)	Returns the cosine of each value.
np9.tan(a)	Returns the tangent of each value.

Example

```
import numpy as np9
import math
a = np9.array([[1,2],[3,4]])
print("np9.exp(a) value :\n ",np9.exp(a))
print("Value of e power 1 ==> ",math.exp(1))
print("Value of e power 2 ==> ",math.exp(2))
print("Value of e power 3 ==> ",math.exp(3))
print("Value of e power 4 ==> ",math.exp(4))
```

OUTPUT

```
C:\h_numpy>py h9.py
np9.exp(a) value :
 [[ 2.71828183  7.3890561 ]
  [20.08553692 54.59815003]]
Value of e power 1 ==> 2.718281828459045
Value of e power 2 ==> 7.38905609893065
Value of e power 3 ==> 20.085536923187668
Value of e power 4 ==> 54.598150033144236
```

Program 1: A program to perform mathematical functions.

```
import numpy as np9
a = np9.arange(5)
print("exp() on a ==> ",np9.exp(a))
print("sqrt() on a ==> ",np9.sqrt(a))
print("sin() on a ==> ", np9.sin(a))
print("cos() on a ==> ",np9.cos(a))
print("tan() on a ==> ",np9.tan(a))
print("log() on a :",np9.log(a))
```

OUTPUT

```
C:\h_numpy>py h9.py
exp() on a ==> [1. 2.71828183 7.3890561 20.08553692 54.59815003]
sqrt() on a ==> [0. 1. 1.41421356 1.73205081 2. ]
sin() on a ==> [0. 0.84147098 0.90929743 0.14112001 -0.7568025 ]
cos() on a ==> [1. 0.54030231 -0.41614684 -0.9899925 -0.65364362]
tan() on a ==> [0. 1.55740772 -2.18503986 -0.14254654 1.15782128]
C:\h_numpy\h9.py:9: RuntimeWarning: divide by zero encountered
in log
  print("log() on a :",np9.log(a))
log() on a : [ -inf 0. 0.69314718 1.09861229 1.38629436]
```

15.2 FINANCIAL FUNCTIONS

NumPy Financial functions are extremely useful and handy for many personal finance calculations, like estimating how much interest you have to pay on a loan and how your money grows with a monthly investment plan with a certain interest rate. Recently, the financial functions are deprecated and now the financial functions are in a standalone NumPy project/package called numpy-financial.

Installation of NumPy Financial

```
>>>pip install numpy-financial
```

(i) fv()

To calculate the future value.

Syntax

```
numpy.fv(rate, nper, pmt, pv, when='end')[source]
```

where

rate	Rate of interest
nper	Number of compounding periods
pmt	Payment
pv	Present value
when	When payments are due ('begin' (1) or 'end' (0)). Defaults to {'end',0}
	{{'begin', 1}, {'end', 0}}, {string, int}

Example: Assume you save ₹200 per month for your future. The average interest rate for your investment is 5%. Now, you are interested to know the value of your fund after 10 years.

The key inputs for **npf.fv()** are as follows:

- Rate of interest as decimal (not per cent) per period (0.5/12)
- Number of compounding periods in months (10*12)
- Payment [(as negative numbers because it indicates that money is going out (–200)]

```
import numpy as np9
import numpy_financial as npf
interest_rate =0.05/12 # 12%
n_periods = 10*12 # 120 months
payment_per_month = -200
present_value =-200
future_value = npf.fv(interest_rate, n_periods, payment_per_month,
                                                present_value)

print(round(future_value))
```

OUTPUT

```
C:\h_numpy>py h9.py
31386
```

Thus, you get ₹31,386 after 10 years.

(ii) pv()

The pv() function is used to compute the present value.

Example: Calculate the present value of an investment that becomes ₹33,139.75 after 10 years on saving ₹200 every month. Assume IR is 5% (annually) compounded monthly.

```
import numpy as np9
import numpy_financial as npf
print(npf.pv(0.06/12, 10*12, -200, 33139.75))
```

OUTPUT

```
C:\h_numpy>py h9.py
-200.0007100715939
```

The present value is ₹200.

(iii) pmt()

The pmt() function is used to compute the payment against loan principal plus interest for the loan.

Syntax

```
Numpy.pmt(rate, nper, pv, fv=0, when='end')
```

Example: For a home mortgage loan, you want to know how much monthly payment you need to make to pay off ₹40,00,000 loan in 30 years at an interest rate of 8%. The pmt() function is used to calculate the payment against loan principal + interest.

```
import numpy as np9
import numpy_financial as npf
interest_rate=0.08/12
```

```
mortgage_amount=4000000
n_periods = 30*12
m_payment = npf.pmt(interest_rate, n_periods, mortgage_amount)
print(round(m_payment))
```

OUTPUT

```
C:\h_numpy>py h9.py
-29351
```

(iv) ppmt()

The ppmt() function is used to calculate the payment against loan principal.

Syntax

```
Numpy.ppmt(rate, per, nper, pv, fv=0.0, when='end')
```

Example

```
import numpy as np9
import numpy_financial as npf
print(npf.ppmt(0.08/12, 1, 30*12, 4000000))
```

OUTPUT

```
C:\h_numpy>py h9.py
-2683.9162885084515
```

(v) ipmt ()

The ipmt() function is used to compute the interest portion of a payment.

Syntax

```
numpy.ipmt(rate, per, nper, pv, fv=0.0, when='end')
```

Example

```
import numpy as np9
import numpy_financial as npf
print(npf.ipmt(0.08/12, 1, 30*12, 4000000))
```
OUTPUT
```
C:\h_numpy>py h9.py
-26666.666666666668
```

(vi) irr()

The irr() function returns the internal rate of return.

Syntax

```
numpy.irr(values)
```

Example

```
import numpy as np9
import numpy_financial as npf
Investment = 500
# Withdrawls at regular interval : 50, 31, 3, 11
print("Internal Rate of Return", npf.irr([-500,50,31,3,11]))
```

OUTPUT

```
C:\h_numpy>py h9.py
Internal rate of return -0.529644772151268
```

(vii) rate()

The rate() function is helpful in computing the rate of interest per period.

Syntax

```
numpy.rate(nper, pmt, pv, fv, when='end', guess=0.1, tol=1e-06, maxiter=100)
```

Example

```
import numpy as np9
import numpy_financial as npf
print(npf.rate(8, 10000, 500, 200))
```

OUTPUT

```
C:\h_numpy>py h9.py
-2.0184122884653117
```

EXERCISES

1. Write a program to compute the future value of an investment.

2. Write a program to calculate the present value of an investment.

3. Explain the role of nper and program it.

16 FUNCTIONAL PROGRAMMING

16.1 INTRODUCTION

We have various functions to perform functional operations on NumPy:

(i) apply_along_axis()

The apply_along_axis() function is used to apply a function to 1-dimensional slices along with the given axis.

Syntax

```
numpy.apply_along_axis(func1d, axis, arr, *args, **kwargs)
```

where

func1d	This function should accept 1-D arrays. It is applied to 1-D slices of arr along the specified axis. Function (M,) -> (Nj...)
axis	Axis along which arr is sliced (integer)
arr	Input array. nd array (Ni..., M, Nk...)
args	Additional arguments to func1d
kwargs	Additional named arguments to func1d (if any)

Example: A program to print average of first and last elements of the 1-D array using apply_along_axis() function.

```
import numpy as np9
def my_func(x):
    return (x[0] + x[-1]) * 0.5 # """Avg first and last element
    of a 1-D array"""
y = np9.array([[12,4,6], [11,3,5], [19,7,8]])
print(y)
print(np9.apply_along_axis(my_func, 0, y))
```

OUTPUT

```
C:\h_numpy>py h9.py
 [[12  4  6]
  [11  3  5]
  [19  7  8]]
  [15.5 5.5 7.]
```

Example: A program to print average of first and last elements of 1-D array using apply_along_axis() function.

```
import numpy as np9
def my_func(x):
    return (x[0] + x[-1]) * 0.5 #"""Avg first and last element
    of a 1-D array"""
y = np9.array([[12,4,6], [11,3,5], [19,7,8]])
print(y)
print(np9.apply_along_axis(my_func, 1, y))
```

OUTPUT

```
C:\h_numpy>py h9.py
 [[12  4  6]
  [11  3  5]
  [19  7  8]]
  [9.  8.  13.5]
```

Example

```
import numpy as np9
y = np9.array([[12,4,6], [11,3,5], [19,7,8]])
print(np9.apply_along_axis(sorted, 1, y))
```

OUTPUT

```
C:\h_numpy>py h9.py
 [[ 4  6 12]
  [ 3  5 11]
  [ 7  8 19]]
```

Example

```
import numpy as np9
y = np9.array([[12,4,6], [11,3,5], [19,7,8]])
print(np9.apply_along_axis(np9.diag, -1, y))
```

OUTPUT

```
C:\h_numpy>py h9.py
 [[[12  0  0]
   [ 0  4  0]
   [ 0  0  6]]

  [[11  0  0]
   [ 0  3  0]
   [ 0  0  5]]

  [[19  0  0]
   [ 0  7  0]
   [ 0  0  8]]]
```

(ii) apply_over_axes() function

The apply_over_axes() function is used to implement a function repeatedly over multiple axes.

Syntax

```
numpy.apply_over_axes(func1, a9, axes1)
```

where

func1   This function must take two arguments, func(a, axis)
a9   Input array
axes1   Axes over which func is applied; the elements must be integers

Example: A program to print an array using apply_over_axes() function.

```
import numpy as np9
x = np9.arange(24).reshape(4,3,2)
print(x)
```

OUTPUT

```
C:\h_numpy>py h9.py
[[[ 0  1]
  [ 2  3]
  [ 4  5]]

 [[ 6  7]
  [ 8  9]
  [10 11]]

 [[12 13]
  [14 15]
  [16 17]]

 [[18 19]
  [20 21]
  [22 23]]]
```

Example

```
import numpy as np9
x = np9.arange(24).reshape(3,2,4)
print(np9.apply_over_axes(np9.sum, x, [0,2]))
```

OUTPUT

```
C:\h_numpy>py h9.py
[[[114]
  [162]]]
```

(iii) vectorize()

The vectorize() function is used to generalize function class.

Syntax

```
class numpy.vectorize(pyfunc1, otypes=None, doc1=None, excluded1=None,
                            cache1=False, signature1=None)
```

where

pyfunc1	A python function or method
otypes	The output data type
doc1	The doc string for the function. If None, the doc string will be the pyfunc.__doc__.
excluded1	Set of strings or integers representing the positional or keyword
cache1	If True, then cache the first function that determines the number of outputs if otypes, is not provided.
Signature1	Generalized universal function signature, e.g., (m,n),(n)->(m) for vectorized matrix-vector multiplication

Example: A program for NumPy.vectorize() method.

```
import numpy as np9
def my_func(x, y):
    if x > y: # Return x-y if x>y, otherwise return x+y
        return x - y
    else:
        return x + y
vec_func = np9.vectorize(my_func)
print(vec_func([12, 14, 16, 18], 4))
```

OUTPUT

```
C:\h_numpy>py h9.py
[ 8 10 12 14]
```

Example: A program on doc string taken from the input function to vectorize unless it is specified.

```
import numpy as np9
def my_func(x, y):
    vec_func.__doc__ # 'Return a-b if x>y, otherwise return x+y'
vec_func = np9.vectorize(my_func, doc="hima 'Educational'")
print(vec_func.__doc__)
```

OUTPUT

```
C:\h_numpy>py h9.py
hima 'Educational'
```

(iii) frompyfunc()

The frompyfunc() function is used to take an arbitrary Python function and returns a NumPy ufunc. It can be used to add broadcasting to a built-in Python function.

Syntax

```
numpy.frompyfunc(func, nin, nout)
```

where

func	An arbitrary Python function
nin	The number of input arguments
nout	The number of objects returned by func

Returns: out : ufunc

Example: A program on NumPy.frompyfunc() method.
```
import numpy as np9
array = np9.frompyfunc(oct, 1, 1)
print(array(np9.array((125, 150, 175))))
```
OUTPUT
```
C:\h_numpy>py h9.py
['0o175' '0o226' '0o257']
```

Example: A program on NumPy.frompyfunc() for comparison.
```
import numpy as np9
array = np9.frompyfunc(oct, 1, 1)
print(np9.array((oct(125), oct(150), oct(175))))
```
OUTPUT
```
C:\h_numpy>py h9.py
['0o175' '0o226' '0o257']
```

(iv) piecewise()

The piecewise() function is used to evaluate a piecewise-defined function. Given a set of conditions and corresponding functions, evaluate each function on the input data wherever its condition is true.

Syntax
```
numpy.piecewise(x, condlist, funclist, *args, **kw)
```
where

x	Input domain
condlist	Each boolean array corresponds to a function in funclist
funclist	Each function is evaluated over x wherever its corresponding condition is True
args	Any further arguments given to piecewise
kw	Keyword

Note: This is identical to select, except that functions are calculated on elements of x that satisfy the corresponding condition from condlist.

Example: A program on NumPy.piecewise() method.
```
import numpy as np9
x = np9.linspace(-3.5, 3.5, 5)
print(np9.piecewise(x, [x < 0, x >= 0], [-1, 1]))
```
OUTPUT
```
C:\h_numpy>py h9.py
[-1. -1. 1. 1. 1.]
```

Example: A program to find the absolute value, which is -x for x <0 and x for x >= 0.

```
import numpy as np9
x = np9.linspace(-3.5, 3.5, 5)
print(np9.piecewise(x, [x < 0, x >= 0], [lambda x: -x, lambda
x: x]))
```

OUTPUT

```
C:\h_numpy>py h9.py
[3.5 1.75 0. 1.75 3.5 ]
```

EXERCISES

1. Explain linspace function with example.

2. Explain piecewise function with suitable example and differentiate it with linspace().

3. What is broadcasting and give an example of frompyfunc().

INDEX